DATE DUE

212 SPECIES IN FULL COLOR

REPTILES AND AMPHIBIANS

by

HERBERT S. ZIM, PH.D., SC.D.

and

HOBART M. SMITH, PH.D.

Illustrated by

JAMES GORDON IRVING

1991 Scholars Edition

GOLDEN PRESS • NEW YORK

Western Publishing Company, Inc.

Racine, Wisconsin

FOREWORD

Reptiles and amphibians constitute a surprisingly diverse and intriguing group (collectively and familiarly called "herps," from "herpetology," the study of amphibians and reptiles). This Golden Guide introduces some of the many adaptations of the fascinating snakes and turtles, frogs and salamanders found in North America.

The authors express their grateful thanks to all who helped in preparing the first edition. Thanks are due to Charles M. Bogert, Bessie M. Hecht, James A. Oliver, Carl F. Kauffeld, Roger and Isabelle Conant, Robert C. Miller, Joseph R. Slevin, Earl S. Herald, L. M. Klauber, C.B. Perkins, C. S. Shaw, Louis W. Ramsey, and William H. Stickel.

Special thanks are due to our colleagues at the University of Illinois—Philip and Dorothy Smith, Harold Kerster, Donald Hoffmeister, and many others. Our gratitude goes also to James Gordon Irving for his fine cooperation and to Grace Crowe Irving; to Rozella Smith and Sonia Bleeker Zim for their assistance; and finally to our publishers for their untiring aid.

The present revision incorporates many changes in nomenclature, classification, common names, and distribution, reflecting the extensive research recorded since publication of the previous edition.

H. S. Z.
H. M. S.

Revised Edition, 1987

 Produced in the U.S.A. Published by Golden Press, New York, N.Y. Library of Congress Catalog Card Number: 61-8324. ISBN 0-307-24057-6

USING THIS BOOK

The first step in the use of this book is to learn the differences between reptiles and amphibians:

TURTLE

REPTILES

Usually: four-legged (except snakes and a few lizards); each foot with three to five clawed toes; skin usually with horny scales, sometimes bony plates. Most lay eggs with hard or leathery skin.

LIZARD

1. TURTLES Leathery or bony shell. Four limbs, short tail. Head can be withdrawn wholly or partly into shell. **pages 18-43**

2. LIZARDS In the United States, mostly small, four-legged, covered with equal-sized horny, smooth or beaded scales. Most are egg-laying, fast-moving land reptiles. **pages 44-69**

SNAKE

ALLIGATOR

3. SNAKES Long, legless. Scales on belly usually larger than others. Skulls loose, mouth large. Lack ear openings. Some are egg-laying; some live-bearing. **pages 70-113**

4. ALLIGATORS and CROCODILES Large, lizardlike. Skull forming long snout. Adapted to water life in warm regions. **pages 114-115**

AMPHIBIANS

Four- or, rarely, two-legged (except tadpoles). Smooth or warty skin, usually moist. No visible scales. Toes never clawed. Eggs usually in jelly-like masses in water.

FROG

1. FROGS and TOADS Adults with much larger hind limbs than forelimbs; tadpoles limbless when young. Adults lack tail. **pages 118-136**

2. SALAMANDERS Most have four limbs, even the larvae. Limbs about same size. Adults have tails. **pages 137-153**

SALAMANDER

The introduction to each of the sections above gives more details. Pages 8-14 explain range maps and suggest activities. Index is on pages 158-160, scientific names on 155-157.

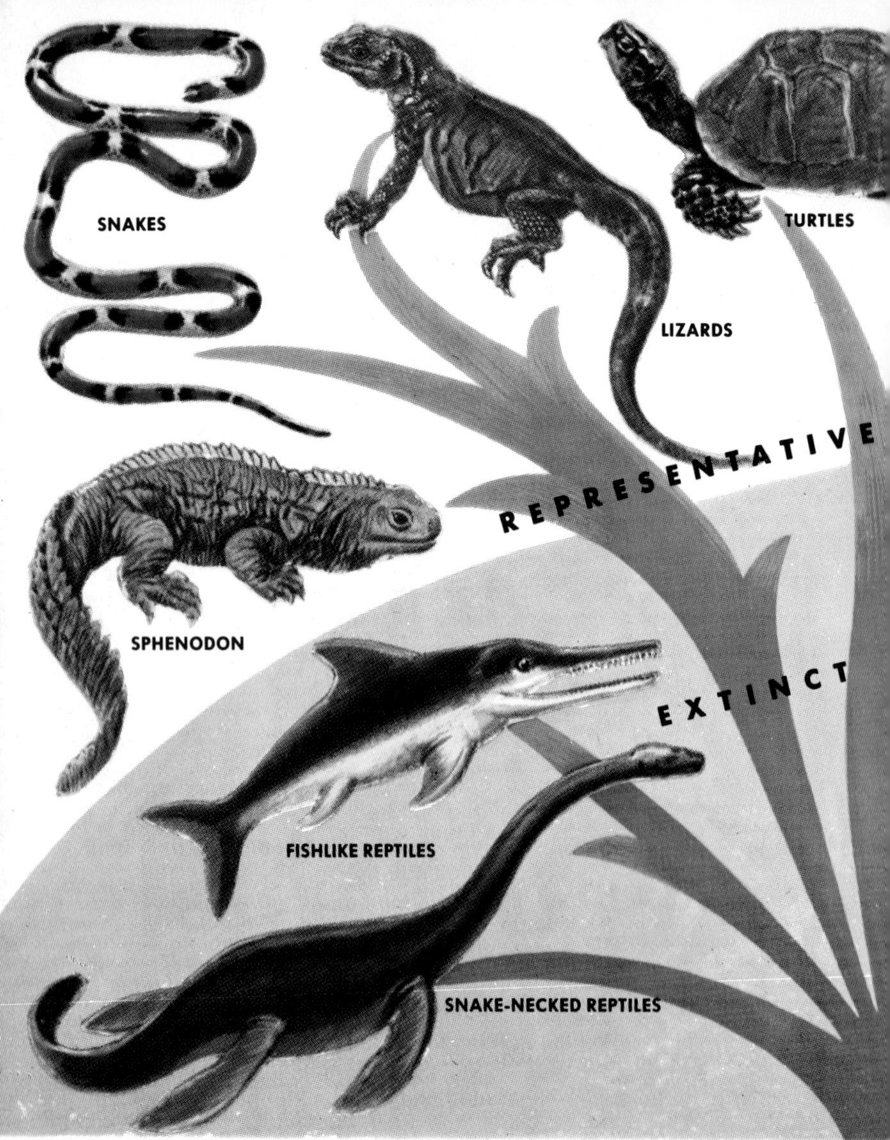

REPTILES appeared nearly 250 million years ago. The group slowly spread, and about 60 million years later took over the land. Dinosaurs included the largest land animals. Other reptiles took to the air and to

the seas. As the climate changed, nearly all the great reptiles died off. Reptiles of today are resourceful descendants of magnificent but adaptively limited ancestors.

AMPHIBIANS had at least a 50-million-year head start on reptiles, but these first land animals never became completely independent of water. Their jellylike eggs could not survive in air, so amphibians had to return

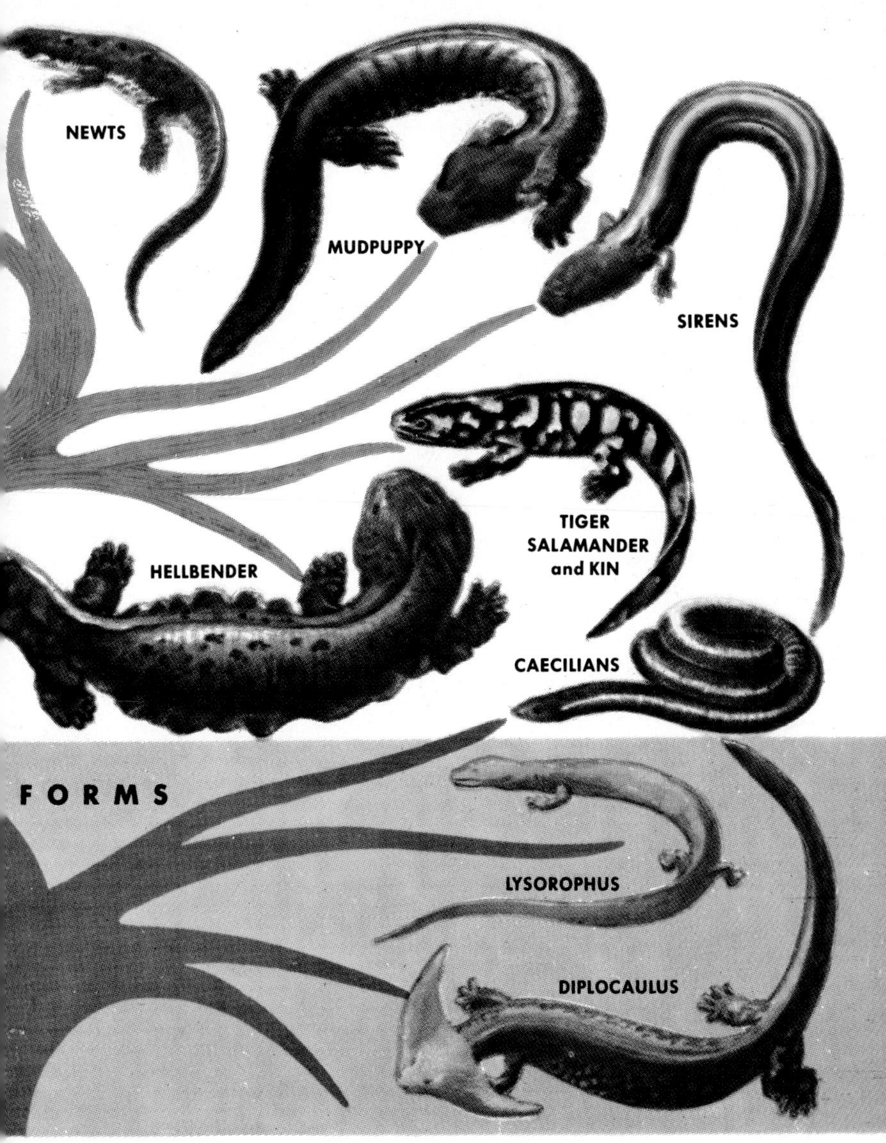

to swamps, ponds, or streams to breed. Ancestors of present-day frogs and salamanders flourished in the Coal Age swamps. Many were clumsy giants. For more about living American amphibians, see pages 116-153.

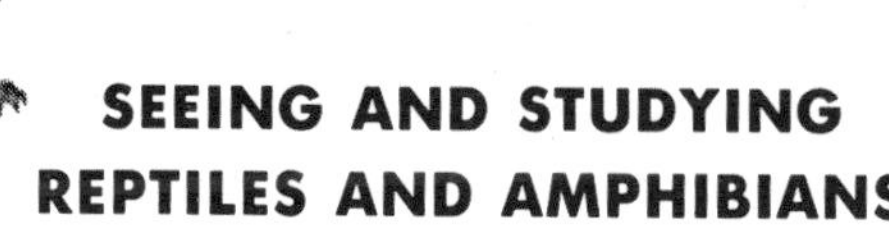

SEEING AND STUDYING REPTILES AND AMPHIBIANS

DESERT IGUANA

COMMON TOAD

MAPS in this book show approximate ranges of our familiar species. Where a map shows ranges of more than one species, the common name of each species is within or next to the color or kind of hatching that shows its range. Overlapping of color and hatching indicates overlapping of ranges.

SCIENTIFIC NAMES of the species illustrated are given on pages 155-157. The scientific name for a species consists of two words—first the generic name (genus) and then the specific name (the two together denoting a species). A third name indicates a subspecies, or race. Knowledge of these names grows increasingly useful as you learn more about reptiles and amphibians.

FACT AND FABLE This book brings together interesting facts and reliable scientific opinions. Sometimes the facts are stranger than fables; sometimes fables are exaggerations or distortions of a small truth. Because some people have mistaken ideas about reptiles and

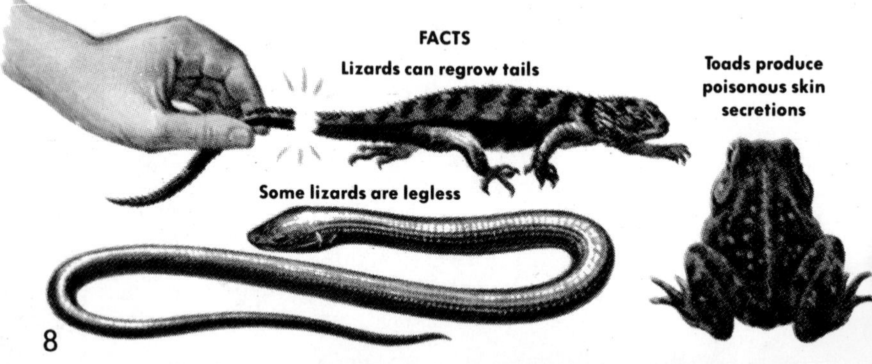

amphibians, they destroy harmless species. We need not fear what we understand; so try to understand these fascinating animals.

INTEREST AND CURIOSITY While some people fear reptiles, most want to see what snakes, lizards, turtles, and frogs are really like. This curiosity has been fed partly by the fables about these creatures and partly by their unusual appearance. Every group of animals includes strange and unusual kinds and, as a group, reptiles and amphibians have the richest share. Reptiles and amphibians have iridescent skin and varied patterns of color that few other animals can equal. They are attractive as well as interesting.

VALUES Some reptiles and amphibians are of direct economic value. In the U.S. most species are protected in most states, but a few common species and captive-bred individuals are subject to limited commercial use and scientific study. Elsewhere many species are becoming rare through over-exploitation. Turtles and their eggs are especially persecuted for food, and snakes, lizards, and crocodilians for leather. Even the venom of poisonous snakes has uses in medicine. In North America, the Gila Monster and poisonous snakes are the only dangerous reptiles. But deaths from snakebite scarcely

total 15 per year in the U.S., far fewer than from insect stings. Reptiles feed partly on rats, mice, gophers, insects, and other pests; in turn they are eaten by mammals and large birds. As a group they play an important role in the balance of nature, increasing in importance toward tropical regions.

Frogs and other amphibians are used in scientific experiments. We eat frogs' legs, and frogs consume quantities of insects. Salamanders, too, serve as food for man and other animals, and help control harmful pests. Like reptiles, amphibians are a vital part of our environment as well as clues to animal life long ago.

CONSERVATION of reptiles and amphibians is important if these animals are to be preserved, for our own enjoyment and for the future. Needless killing, often based on fear and misunderstanding, should stop. No reptiles should be killed, except poisonous snakes near human habitations. Even more important is the preservation of wild areas where reptiles, amphibians, and other wildlife live. The cutting of forests, draining of swamps, damming of rivers, and building of roads in wilderness areas all have a long-range effect on plant and animal life. The preservation of unspoiled land in state and national parks and forests, wildlife refuges, and wilderness areas is important in the conservation program. Swamps and marshes on farmland are worth keeping too.

Many reptiles die as an indirect result of their being cold-blooded. Snakes often come out on roads at night, possibly because of the warmth of the pavement. Turtles are constantly crossing highways, too. An early morning ride will show the toll taken by passing cars—a toll that could be reduced by more care taken on the part of motorists.

LEARN TO KNOW reptiles and amphibians from books and, better, from life. Learn those in your region first. Be able to recognize poisonous snakes at a glance. Besides the zoo, make use of museums and exhibits. Become familiar enough with lizards, turtles, snakes, frogs, and salamanders to recognize common ones seen in the field.

FIELD STUDIES in your own region come next. If you can, go with an experienced person. Hikes will make you familiar with places where reptiles and amphibians are found. The best places depend largely on local conditions. Go to ponds and swamps, creeks, ledges, woods, and fields. This is the first step in observing or collecting.

COLLECTING may seem more important than it is. You can learn much without collecting. Only for the advanced student is it necessary to pin down rare species or geographical subspecies for study of body scales, head plates, forms of toes, and other details. At that stage systematic collecting is important. If you collect harmless species, turn them loose after you have examined and studied them.

Hunting Frogs with Headlight

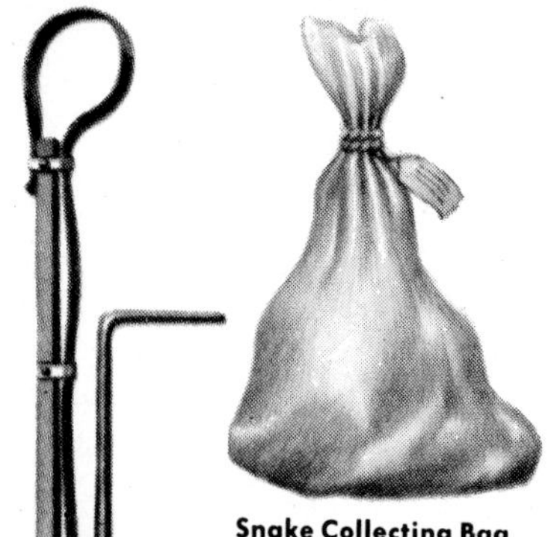
Snake Stick and Noose

Snake Collecting Bag

COLLECTING EQUIPMENT need not be fancy. Snakes and lizards can be carried in muslin bags or pillow cases. When the end is tied, these are safe, and provide enough ventilation. Cans or plastic (not glass) jars are also fine for amphibians. Keep moist leaves or grass in the container to prevent drying out. A stout net will help you get amphibians, though some collectors prefer to grab by hand. A snake stick will pin down a snake till you can pick it up safely. Some prefer to grab snakes quickly behind the head. Using a stick is safest for a beginner, however. Amateurs should leave poisonous snakes strictly alone. Experience in field trips will help you plan simple but adequate collecting equipment and the proper ways of using such equipment. Remember that a muslin bag can be a death trap for a specimen if left in the hot sun or in a closed-up parked car.

Terrarium for Frogs

Cage for Snake or Lizard

CAGES AND TANKS Keep amphibians in aquaria. Some require a rock or float so they can climb out of the water. Others, especially tadpoles, will use any aquaria suitable for fish. Toads will need a moist terrarium; lizards use a cage similar to one for snakes. For snakes and lizards that climb, use a larger cage with a branch set in it. Allow at least a square foot of floor space per foot of a medium-sized snake, more for larger species. Know the habits of the snake. Try, in a simple way, to duplicate the natural habitat. A wooden cage of one-inch boards with a glass front is good; the top should be hinged and used as a door. Three or four one-inch (or larger) holes at the ends and back aid ventilation. These holes should be tightly covered with fine screen. Put sand, gravel, or newspaper on the floor. Add a rock or two and a large enough dish of clean water so that your snake or toad can soak. Be sure that the floor of your cage is always dry. Fasten your water container so a moving snake will not turn it over. Reptiles kept in wet cages often develop skin infections which are difficult to cure. Turn such sick snakes loose.

Amphibian Eggs in Aquarium

LIFE HISTORIES of many amphibians and reptiles are still unknown. Sometimes only the adults have been described, and we know nothing of their eggs or young. Eating habits, wintering habits, and mating habits of many species are still mysteries. A careful, informed student may be able to make accurate field observations and records of scientific value. Binoculars are often a help, and a notebook is essential. Field observation may teach you much more than watching animals in a cage. For best results, combine both methods. First study the animals carefully in the field. Then observe them in captivity for further details. The more natural the conditions, the better the observations.

REPTILES AND AMPHIBIANS IN CAPTIVITY Although it is easy to keep reptiles and amphibians in captivity, since their needs are so simple, many species cannot legally be possessed without state or federal permit. Inquire at your state Wildlife Department and consult local herpetologists before you attempt to keep any wild-caught reptile or amphibian, or even to kill poisonous species. Animals caught should be returned to where they were found as soon as possible after you have studied and photographed them. Many species, usually needing no permit to keep, are available from commercial dealers. Some cities ban all snakes as captives, and poisonous species should never be kept.

Holding Snake Safely

FIRST AID FOR SNAKE BITE

Attempts at first aid often do more harm than good. If given at all, it should be at once.

Read this *before* you begin handling snakes.

Snakebites of any kind are rare. They are easily prevented. Wear heavy shoes, boots, or leggings in country where poisonous species are found. Stay on roads, paths, or trails if possible. Step clear of rocks and logs. When climbing rocky ledges, look before you grasp. Finally and most important: no amateur should catch or handle poisonous snakes.

Bites of non-poisonous snakes often leave a U-shaped pattern of tooth marks. Treat them as simple, minor wounds with any good germicide. Bites of poisonous snakes usually show a double puncture caused by the enlarged front fangs. Other teeth marks may be present also. If bitten, try hard to recognize and identify the snake. If the snake is poisonous, complete quiet with prompt first aid, and the use of serum by a doctor, will insure the possibility of complete and rapid recovery.

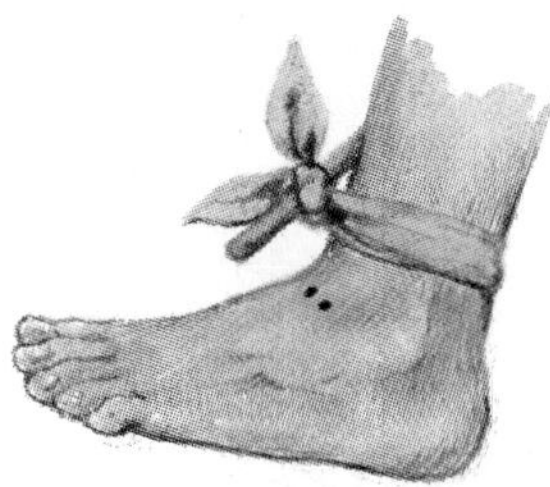

Apply constriction band, not a tourniquet.

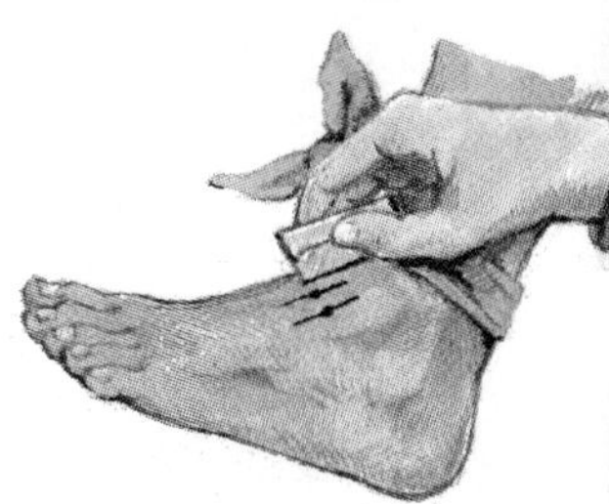

Make at once a shallow, lengthwise incision at each fang entry, within seconds if possible.

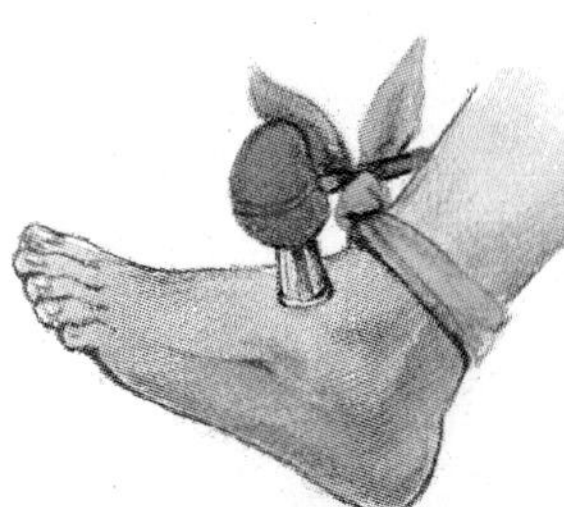

Suck out poison.

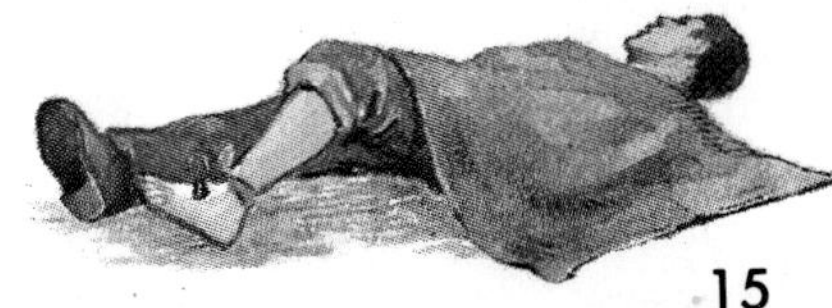

Keep patient quiet, warm, and comfortable.

Phone or transport to doctor immediately.

TURTLES
(49 U.S. Species)

LIZARDS (95 species)

Scaly Skin

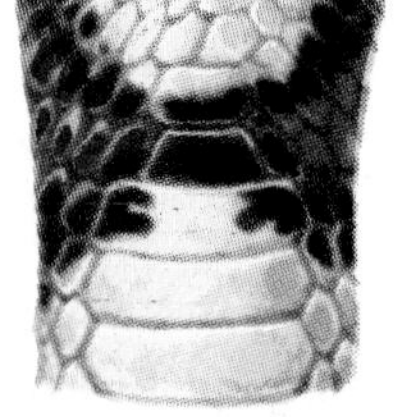

Plated Skin

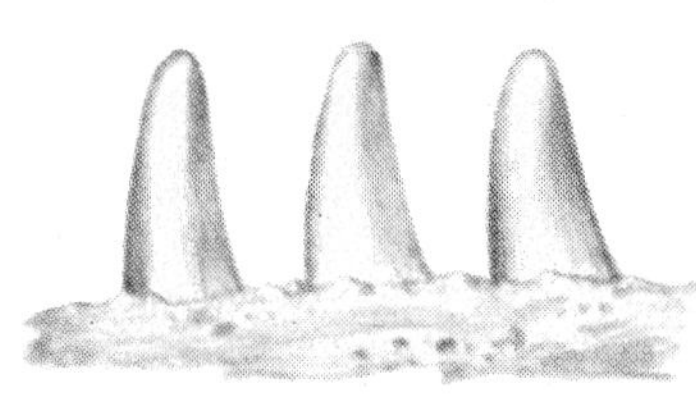

REPTILES (teeth alike)

REPTILES Though the Age of Reptiles, which flourished for some 120 million years, came to an end about 70 million years ago, many interesting and unusual reptiles are still found today. Some native reptiles occur in every state, though they are more common and more species occur in the warmer parts of this country. Reptiles are classified into four major groups—turtles (49 species), lizards (95 species), snakes (120 species), and alligators and crocodiles (2 species). Reptiles are not always easy to find. Some are small, many are nocturnal, and most are protectively colored.

Reptiles are cold-blooded. A reptile's body temperature is the same as the temperature of its surroundings, except as evaporation lowers it or the sun's warmth raises it. Only by behavior is a reptile's temperature controlled. In the sun or on hot ground a reptile can get much warmer than the air. Desert reptiles avoid direct

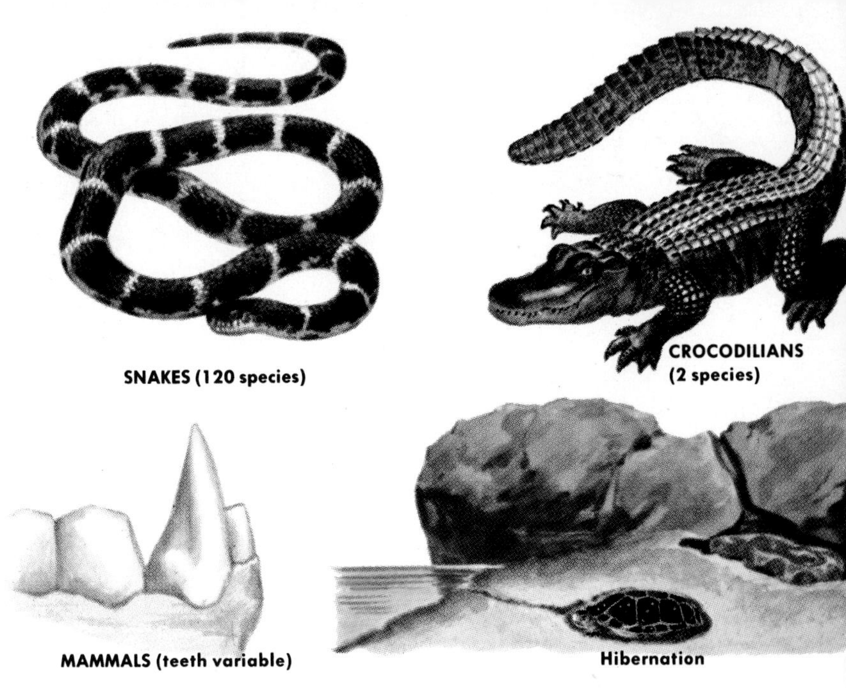
SNAKES (120 species)

CROCODILIANS (2 species)

MAMMALS (teeth variable)

Hibernation

midday sun. Some become dormant (aestivate) in midsummer. In cooler regions reptiles hibernate from late fall to early spring under soil, rocks, or water. Then they are inactive, sometimes almost lifeless.

All reptiles, even aquatic species, have lungs and breathe air. Their skin is usually covered with scales or plates. Reptilian teeth are commonly uniform in shape and size, lacking the specialization seen in mammals. Most reptiles lay eggs. In a few, the eggs develop inside the mother and the young are born alive. All young are able to care for themselves very soon after birth or hatching. Though a few snakes and lizards are poisonous, the great majority of reptiles are harmless. Almost all are predominantly beneficial to man because they feed on rodents and insects. Their unusual habits and structure make them fascinating to study. All deserve protection from needless destruction.

BOX TURTLE

TURTLES are unusual, ancient reptiles. Their ancestors first appeared some 200 million years ago, long before the dinosaurs. And while those great beasts have long been extinct, turtles with their odd, ungainly form have managed to survive and have remained relatively unchanged for at least 150 million years. Part of the reason for this long survival may be the turtle's unusual skeleton. The top shell or carapace is formed upon overgrown, widened ribs. Beneath is the lower shell, or plastron. In the course of their development, turtles have become so modified that their legs are attached within their ribs. This development for protection has made it necessary for turtles to develop long necks and an unusual way of getting air in and out of their lungs. A turtle's neck forms a tight S-shaped bend, and the curve becomes shallow as the neck extends.

Turtles have no teeth. But their horny bill scissors plant and animal food. Turtles eat insects, worms, grubs, shellfish, fish, and some plants. A few species are largely

Eggs SLIDER Young

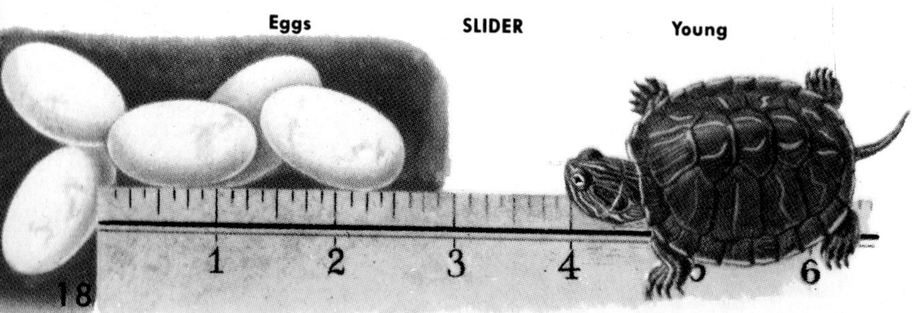

herbivorous. All turtles lay eggs, usually 6 to 12, and bury them in the ground. Sea Turtles lay many more. Under the heat of the sun these hatch into young which grow to maturity in about 5 to 7 years. Turtles may live longer than any other animals, certainly up to 150 years. Small species have survived longer than 40 years in captivity.

Male turtles are generally smaller than females; they have a longer tail, often a concave plastron, and long nails on their front feet. In many kinds, sex is determined by the temperature at which the eggs develop. In northern sections of the country, turtles hibernate under soil or under mud at the bottom of ponds. Some also become dormant in hot, dry weather. Several kinds are prized as food.

Living turtles of North America and adjacent seas fit into seven families. Six are illustrated at the right side of the page by representative species. A representative of the seventh family, the land tortoises, is pictured on page 27. This is the only group correctly called "tortoises."

LEATHERBACK

GREEN TURTLE

MUD TURTLE

SNAPPING TURTLE

SOFTSHELL TURTLE

FALSE MAP TURTLE

SEA TURTLES are larger than, and different from, pond and land species. The limbs of marine turtles are modified into flippers—streamlined for swimming, clumsy for use on land. As a result, these turtles seldom come ashore, though the female does so to lay her large batch of eggs in late spring. The eggs are buried in the sand just past the high-water mark. Sea Turtles are found in warmer waters of both Atlantic and Pacific, and occasionally off northern shores in summer. Of five kinds, the Leatherback is largest. Specimens over 8 ft. long, close to 1,500 lb., have been

caught. The ridged, leathery back makes identification easy. The Hawksbill, smallest of the Sea Turtles, also is easy to recognize because of its overlapping scales. This is the species from which "tortoise shell" comes. The Green Turtle has four plates on each side between the top and the marginal plates. The Loggerhead and Ridleys (not illustrated; two species) have five to seven plates on each side. The Loggerhead is much larger than the Ridleys and usually has three scales at the sides of the plastron; the Ridleys have four.

MUSK TURTLES are aquatic species of ponds, slow streams, and rivers. They often sun themselves in shallow water, but seldom come ashore. The females do so to lay eggs. Note the narrow, high carapace, often covered with algae and water moss. The lower shell (plastron) is narrow and short, almost like that of Snapping Turtles. The Musk Turtle has a strong odor. Four species occur; the commonest, shown above, has two light stripes on each side of its head.

EASTERN MUD TURTLE

YELLOW MUD TURTLE

MUD TURTLES, five species of them, live about the same as Musk Turtles. They are aquatic, feeding on larvae of water insects and small water animals. Notice that the plastron is much wider in the Mud Turtle and is all scaly. Both ends are hinged, so that the Mud Turtle can pull the plastron in, giving head and limbs more protection. Mud Turtles have a musky odor, too. They are small, rarely over 4 in. long, and are more common in the Southeast.

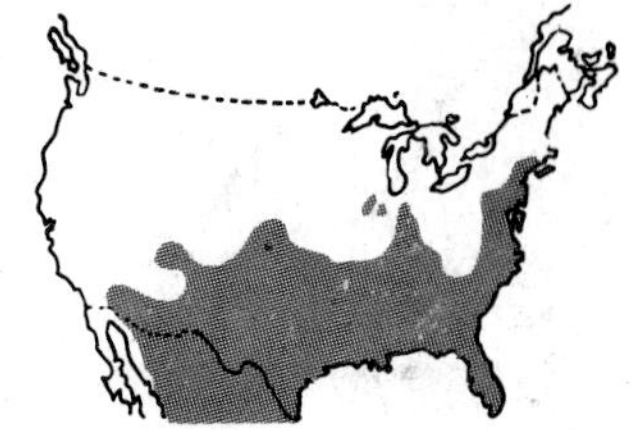

SNAPPING TURTLE and its giant relative (p. 25) are dangerous. Their long necks, powerful jaws, and vicious tempers make them unsafe to handle. Experts carry them by the tail, well away from the body. Snappers are aquatic, preferring quiet, muddy water. They eat fish and sometimes waterfowl. Note sharply toothed rear edge of the rough carapace, often coated with green algae. Plastron is small. Adults, to 18 in. or more, weigh 20 to 62 lb (86 in captivity).

ALLIGATOR SNAPPING TURTLE is the largest freshwater turtle, reaching a length of 30 in. and a weight of close to 235 lb. Entirely aquatic, it lies on the muddy bottom, its huge mouth agape, wiggling a pink, wormlike growth on the tongue to attract unwary fish. Its powerful jaws can maim a hand or foot. It differs from Common Snapper in having three high ridges or keels on its back. Specimens are reported to have lived 60 years and more in zoos.

SPINY SOFTSHELL TURTLE

SOFTSHELL TURTLES have, in fact, a hard shell, but it is soft-edged and lacks horny scales. These turtles can pull in head and limbs for protection nevertheless. Of three species, two have small bumps or tubercles along the front edge of the carapace; the other does not. All have long necks, sharp beaks, vicious tempers. Handle them by rear of shell. These turtles grow to a length of about 20 in. and weigh up to 35 lb. They are excellent eating.

GOPHER TORTOISE

TORTOISES are land turtles with blunt, club-shaped feet very different from the webbed feet of aquatic species. Their diet includes much plant material as well as insects and small animals. Our three species differ externally in minor ways but are placed in two genera. Their family includes the Giant Tortoises of the Galapagos Islands, largest of land turtles. The high, arched carapace and the habit of burrowing are characteristic except in the Texas species.

BASKERS are a common group of six species. The carapace is usually smooth and fairly flat, the rear edge roughly toothed. The carapace of the Florida and Alabama species arches higher than the carapace of others. The olive-brown shells and skins of Baskers are splotched with red and yellow. The Red-eared Slider has a distinctive dash of red behind the eye. The males, much darker than females, were once mistaken for different species. With the extra-long toenails on their front feet they seem to tickle or gently scratch the female's head during

courtship. The female later digs a hole near the shore and deposits about 10 eggs, which she covers with dirt.

All Baskers prefer the quiet waters of rivers and ponds. On warm days they may be found sunning on logs or debris. They are one of the commonest turtles of the Mississippi and its tributaries. Of all young turtles sold in pet shops, Baskers are commonest. They are omnivorous, live long, and grow to about 1 ft. Painting their shells deforms and may finally kill them.

RIVER COOTER used to be called the Hieroglyphic Turtle because the markings on its shell and skin resemble ancient writing. It is a typical Basker with a dark, flattened carapace, 10 to 12 in. long, marked with yellow. The plastron is yellow with dark markings. Like other Basking Turtles, this one feeds on small water animals, insects, and even dead fish; it also eats some water plants. In the various parts of the South, Baskers are prized for their flavor.

CHICKEN TURTLE is so called because it is locally eaten despite its size (5 to 8 in.). The brownish carapace has shallow furrows, a smooth rear edge, narrow yellow lines, and is higher and narrower than in the Baskers. The plastron is yellow, as are the undersides of head and limbs, which have fine, dark stripes. Chicken Turtles have very long necks. They prefer ditches and ponds to rivers. More pugnacious than Basking Turtles, they should be handled with care.

EASTERN PAINTED TURTLE

PAINTED TURTLES are perhaps the most common and widespread of turtles. They are found wherever there are ponds, swamps, ditches, or slow streams. These small (5 to 6 in.) turtles spend much of their time in or near water, feeding on water plants, insects, and other small animals. They are also scavengers. In summer, Painted Turtles gather together, and if one approaches quietly, they may be seen sunning on logs, rocks, or even floating water plants. Males are similar

SOUTHERN PAINTED TURTLE WESTERN PAINTED TURTLE

MIDLAND PAINTED TURTLE

to the females but smaller, with the same long nails on their forefeet that Baskers have. Females lay 6 to 12 white eggs in a hole that they have dug laboriously with their hind legs in the soil. The eggs may hatch in two or three months, though some young do not emerge till the following spring. Painted Turtles may be easily identified by their broad, dark, flattened, smooth-edged shells. The margin of the carapace is marked with red; so is the yellow-streaked skin, especially on head and limbs. The plastron is yellow, sometimes being tinted with red. In all four subspecies of Painted Turtles the upper jaw is notched in front. The notch has a small projection on each side. Markings and details of carapace and plastron differ from subspecies to subspecies. Painted Turtles are shy and are not easily captured. Hardy and adaptable, they survive well even in urban areas and in harsh cold. Though frantic when first captured, the turtles are not pugnacious.

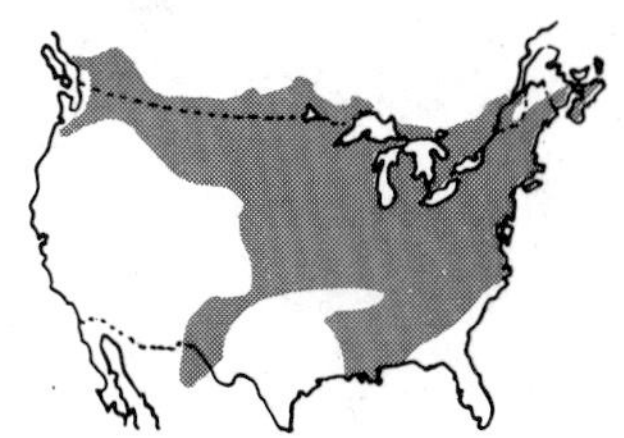

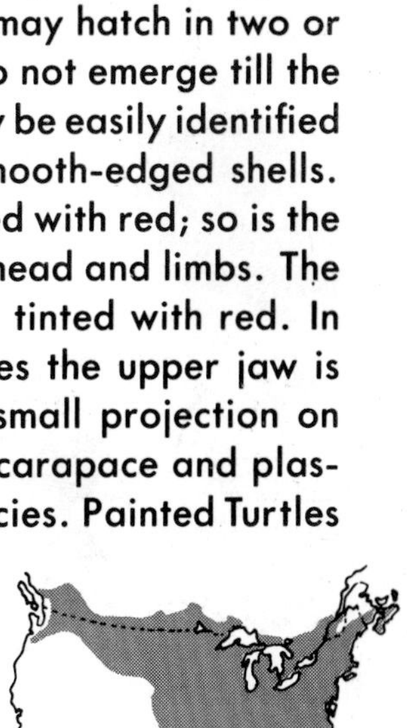

MAP TURTLES are aquatic turtles often found in large numbers in ponds, swamps, and quiet streams. They are even more timid than Painted Turtles. Differences between the sexes are more extreme than in other turtles, females reaching nearly 13 in., males less than half that. Females develop a grotesquely broad head with massive crushing jaws, adapted for feeding on hard-shelled clams and snails. Males and juveniles feed on soft-bodied insects and other aquatic animals. Males seek shallow, debris-laden waters, often sunning them-

MAP TURTLE

selves; females remain mostly in open, deep, muddy-bottomed waters and seldom sun themselves. The female, coming ashore briefly in early summer to lay 10 to 16 eggs, returns to the water as soon as the eggs are buried. Map Turtles (ten species) are named for the faint yellow pattern on the carapace. Lines are brighter on head and limbs. The keeled carapace and its roughly toothed rear edge are identification marks.

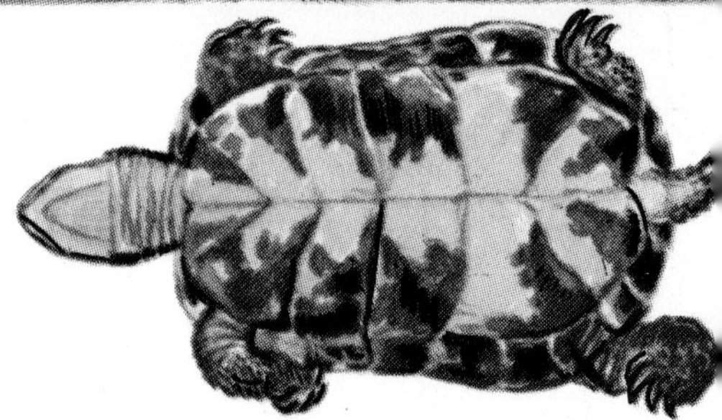

BLANDING'S TURTLE with its hinged plastron somewhat resembles the Box Turtle, but cannot close its shell tightly. It has webbed feet and lacks the hooked bill of the Box Turtle. The plastron is notched at the back. Reaches a length of 10¼ in., but is commonly 7 to 8 in. Prefers quiet waters, but also lives in marshes, where it feeds almost entirely on crayfish and insects. Yellow and black markings make this shy species especially attractive. It is becoming rare.

TERRAPIN, often called Diamondback because of the angular rings on the carapace plate, is the best-known eating turtle. It used to be raised on turtle farms, 8-in. specimens selling for as high as $10, but in recent decades it has lost much of its popularity. Young are protected by law. These turtles of brackish water and tidewater streams have webbed feet. The plastron is yellow. Females are larger. Food: small shellfish, crabs, worms, plants.

EASTERN BOX TURTLE

BOX TURTLES are land species, occasionally found in or near water, though they are well adapted for life on land. They prefer moist, open woods or swamps and feed on insects, earthworms, snails, fruits, and berries. Box Turtles have a hinged plastron which they pull tight against the carapace for complete protection when they are frightened. The carapace, 4 to 5 in. long, is highly arched. Of the two species of Box Turtles, Eastern and Western, the former is divided into four subspecies, the latter into two, distinguished by the shape and markings on the shells and by the number

EASTERN BOX TURTLE

WESTERN BOX TURTLE

WESTERN BOX TURTLE

of toes (three or four) on the hind feet. The plastron of the female is usually flat; that of the male, curved inward. Males have longer tails, and the eye of the male is usually bright red. The female has dark reddish or brown eyes. In early summer the female buries four or five round, white eggs in a sunny spot. These hatch in about three months. The young may hibernate soon after, without feeding. Young Box Turtles grow ½ to ¾ in. yearly for five or six years; then they grow slower—about ¼ in. a year. At 5 years they mate and lay eggs; at 20 they are full-grown, and they may live to be as old as 80. Box Turtles have been reported living 25 years and more in captivity. Though seemingly docile, some bite unpredictably. They eat poisonous mushrooms without harm, but the poison is stored in their flesh, which when eaten by other animals (including humans) has caused death.

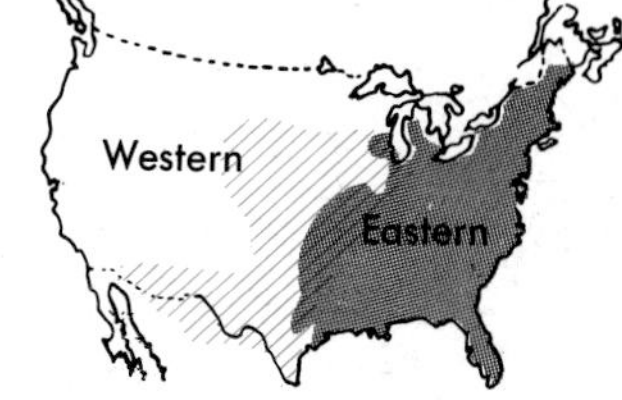

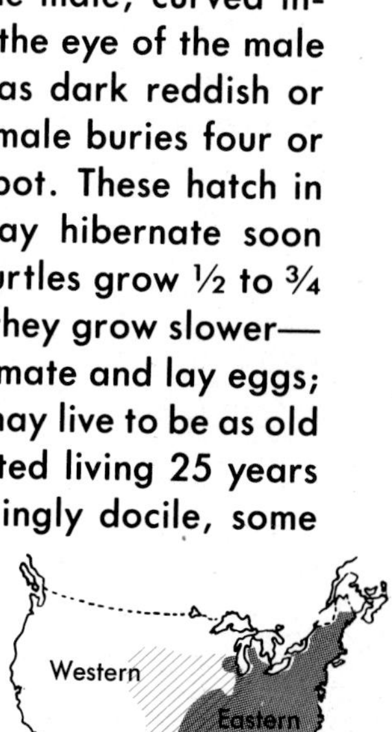

SPOTTED TURTLE is a small (3 to 5 in.) common spring turtle with round orange or yellow spots on its smooth, black carapace. The head is colored similarly. The young have but one spot on each scale, or none. Living in quiet fresh water, this turtle feeds on aquatic insects, tadpoles, and dead fish, but eats only when in water. The tail of the male is about twice as long as the female's. Usually three eggs are laid in June.

WESTERN POND TURTLE is related and similar to the eastern Spotted Turtle, but larger—6 to 7 in. The yellow dots and streaks on the carapace are faint. The plastron, concave on the male, is yellow with dark patches at the edges. This is the only fresh-water turtle of the far West. Living in mountain lakes, marshes, and in slow stretches of streams with abundant aquatic vegetation, Western Pond Turtles feed on small water life, including some plants.

BOG TURTLE, smallest in the world, is quickly identified by large orange spot on each side of head. Dark carapace is short (3 to 4 in.) and narrow, marked with concentric rings. This turtle is semiaquatic, living in mud-bottomed bogs, swamps, and slow streams, feeding omnivorously. In June-July, three to five eggs are laid. Once popular as a pet species, it is now federally protected throughout its range; state regulations also limit possession without a permit.

WOOD TURTLE is recognized easily by its bright, orange-red skin and its heavy, keeled carapace with deep concentric grooves. Adults are 7 to 9 in. long. They prefer moist woods, though they move into open land to feed and, when it is dry, to swamps and into ponds and slow streams. They are omnivorous, nest in May-June, and lay 4 to 12 eggs. The species is protected throughout its range. Male has heavier, longer claws, and larger plates on its forelimbs.

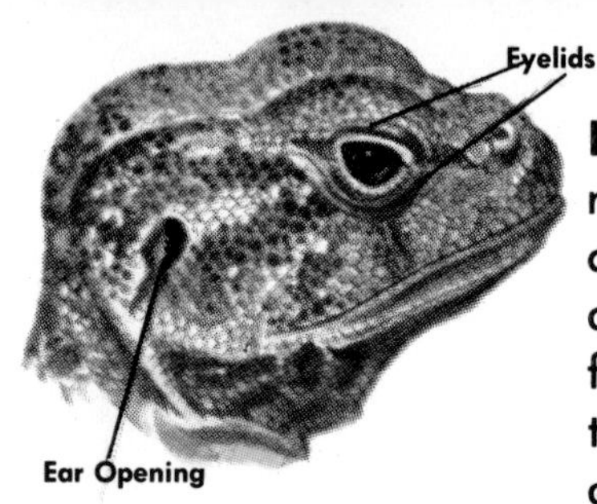

LIZARDS are more like ancient reptiles than snakes (their derivatives) are. Bones of fossil lizards have been found in rocks formed during the period when the dinosaurs were common. Lizards are found mainly in the warmer parts of the world, though a few species live as far north as Canada and Finland. Over 2,500 species are known. These have been grouped into about 20 families, 9 of which occur in the United States. About 350 species are found in North America; 95 of these occur within the boundaries of the United States.

A few species of lizards (pp. 67-68) are snakelike in appearance; they have long bodies and have lost all traces of legs. In all other characteristics, however, they resemble other normal lizards, and close observation easily distinguishes them from snakes. Lizards are typically four-legged, with five toes on each foot. They have scaly skins. On the underside the scales form several or many rows, in contrast to most snakes, which have only a single row of scales.

Salamanders (pp. 137-153) are sometimes mistaken for lizards. Salamanders have smooth skins, live in moist places, have less than five toes on the front feet.

Lizards usually have movable eyelids and an ear opening on the side of the head (snakes have neither).

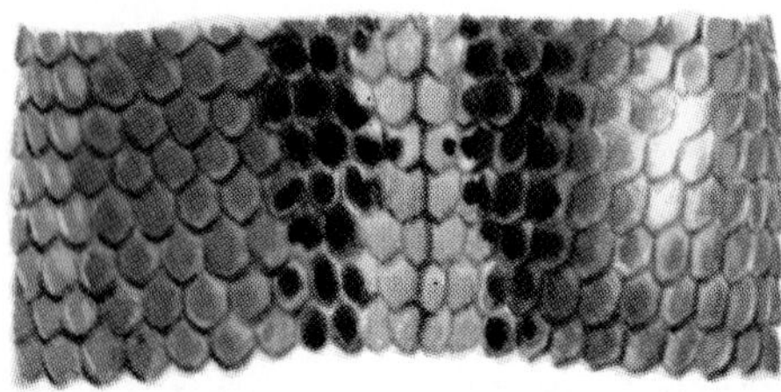

Belly Scales—LIZARDS

Belly Scales—SNAKES

Most lay eggs, but in a few kinds the eggs develop inside the mother's body and the young are born alive. The males and females are alike in many species; in other species they are different in size and color. Many lizards feed on insects and other small animals, such as those illustrated on this page, but a few species feed on plant material. They recognize their prey by its movement and grasp it with lightninglike speed. Lizards can run rapidly—the fastest has been clocked at about 15 miles per hour. Most can swim. Some desert species move through the sand just below the surface with a swimming movement.

Lizards are not easily caught, but those that are can be interesting to observe. Leave the venomous Gila Monster (p. 69) alone, even though it may not be so dangerous as poisonous snakes. Males of given species are intolerant of each other, especially in the breeding season. Females are less aggressive. Some kinds live up to 54 years, others average less than a year, in nature.

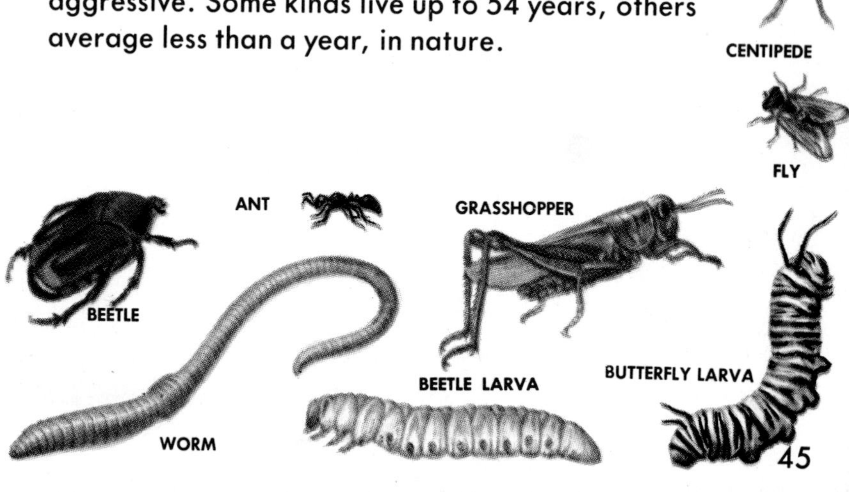

LEAF-TOED GECKO

GECKOS are unusually attractive lizards, recognized by their large, often lidless eyes with vertical pupils. The skin, usually covered with fine, beaded scales, is almost transparent. Most have enlarged, padded toes. Some live around houses, others on rocks or trees, feeding on small insects. They are nocturnal. Geckos lay one to three or more small white eggs with brittle shells during summer. Five or more tropical species have become naturalized in Florida. Most are docile and rarely bite. The tail breaks off easily. Leaf-toed and Banded Geckos are our only truly native species. The Dwarf Gecko long ago came to Florida from the West Indies, the Turkish Gecko recently from North Africa.

DWARF GECKO

TURKISH GECKO

BANDED GECKOS (four similar species) are western lizards (4 to 5 in. long). Tails are about half the body length but have usually broken and regrown shorter. Banded Geckos live in rocky or sandy deserts and lower mountain slopes. They come out at night to feed on small insects and, in turn, are eaten by snakes and larger lizards. They hibernate from Oct. to May but are common other months. They never bite, but are shy, quick, and easily injured.

Banded
Other Geckos
Leaf-toed

GREEN ANOLE (male)

GREEN ANOLE (female)

TRUE CHAMELEON

GREEN ANOLE, common and attractive, can change color, often matching its background of leaves or branches. This is the "chameleon" sold at fairs and circuses. Males have a flap of skin on their throat. Eggs are laid singly every two weeks in summer under moist debris; they hatch in about six weeks. At least four other species of anoles have become established in Florida, from the West Indies. True chameleons are Old World lizards.

CHUCKWALLA (16 in. long) is, next to the Gila Monster (p. 69), our largest lizard. It feeds on flowers and fruit of cactus and tender parts of desert plants, and usually eats well in captivity. Chuckwallas sun themselves on rocks but, when disturbed, dart into crevices, where they inflate their bodies and are difficult to remove. Indians used to eat them. The young have bands across body and tail; the adults have tail bands only.

DESERT IGUANA or CRESTED LIZARD, a handsome spotted species of open deserts, lives in burrows under sparse shrubs. Entirely vegetarian, it feeds on tender desert plants. Desert Iguana is fairly large (12 to 15 in.), but its tail is almost twice its body length. It runs rapidly, is wary and hard to catch. Each has its own territory for feeding where the female lays her eggs. Males have a reddish patch on each side of the tail.

SPINY and TRUE IGUANAS, representing two groups of large American lizards, are not found natively in the U.S., but come to within 100 miles of our border and have established colonies through introduction in Florida, Texas, and California. About 10 or 11 species of Spiny or False Iguana (1 to 4 ft. long) live in Mexico and Central America. The True Iguana (4 to 6 ft.) lives in the same area, mostly in trees. Both are locally favorite foods. Both are herbivorous. Other Iguanas live in the Galapagos Islands and West Indies.

COLLARED LIZARDS The black collar marks the Collared Lizard (four species); so does its long tail, plump body, thin neck, and relatively large head. Males are more brightly colored, with a tinge of orange and yellow. Collared Lizards, fairly common in rocky areas, feed on insects and small lizards. Wary, they can run swiftly on their hind legs. Collared Lizards bite when captured, lay 1 to 2 clutches of 1 to 13 eggs. A species of the lower Rio Grande valley lacks the black collar. A cornered animal gapes the mouth widely, showing its jet black lining.

LEOPARD LIZARD

LEOPARD LIZARDS (two species), somewhat like Collared Lizards in form and size, are more spotted, with narrower heads and bodies. Prefer flat, sandy areas with some vegetation; eat insects and lizards, and often its own kind. Oddly, females develop a deep salmon color on their undersides at close of the breeding season. They lay 1 to 2 clutches of 1 to 11 eggs, which hatch in a month or so. San Joaquin Valley species is endangered. Both species aggressive, bite freely.

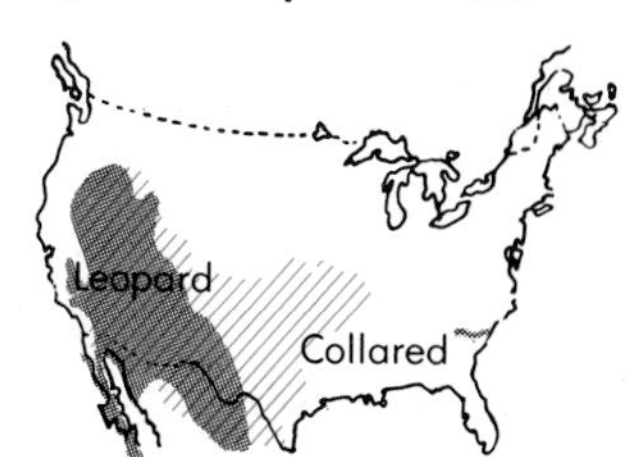

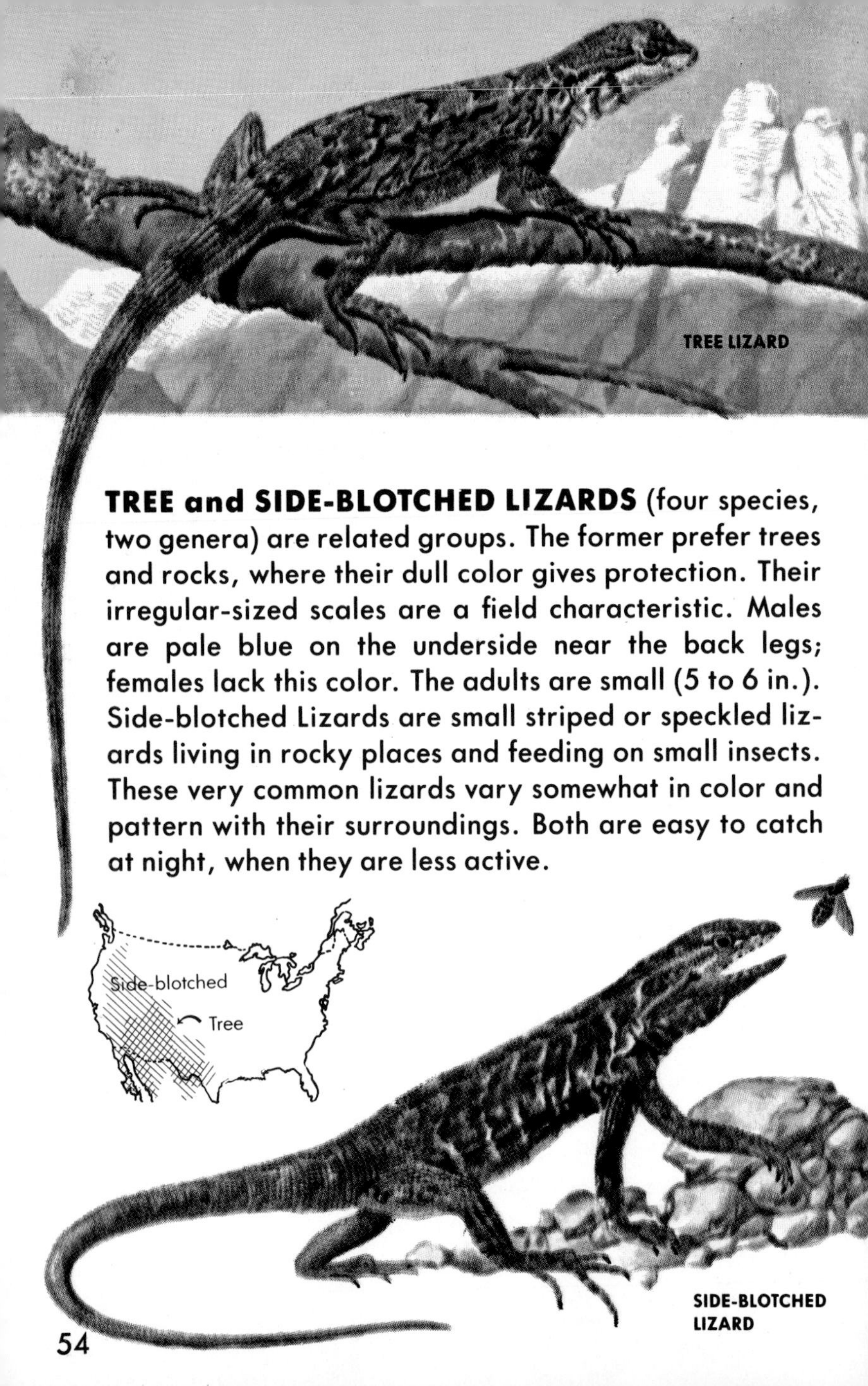

TREE and SIDE-BLOTCHED LIZARDS (four species, two genera) are related groups. The former prefer trees and rocks, where their dull color gives protection. Their irregular-sized scales are a field characteristic. Males are pale blue on the underside near the back legs; females lack this color. The adults are small (5 to 6 in.). Side-blotched Lizards are small striped or speckled lizards living in rocky places and feeding on small insects. These very common lizards vary somewhat in color and pattern with their surroundings. Both are easy to catch at night, when they are less active.

SAND LIZARDS include eight medium-sized (6 to 8 in.) lizards (of four genera), all preferring sandy terrain. Two of the species above are at their best in the sand dunes of the California and Arizona deserts. All have a skin fold across the underside, in front of the forelegs. Legs and fringed toes are long. Tails, about body length, are often marked with black bars underneath. Lip scales slant, help in sand-burrowing. All feed on small insects.

Zebratail
Lesser Earless
Fringe-toed

SPINY LIZARDS form a large group of common lizards, including Fence, Spiny, and Sagebrush Lizards. Some 37 forms (species and subspecies) live in the U.S., almost three times as many farther south. The largest have bodies about 5 in. long, tails slightly longer. All are active in daylight, spending the night in cracks, crevices, or on branches. Some species lay eggs; others bear 6 to 12 young alive. Head, body, and limb forms are guides to the entire group, once you learn them. These lizards lack the skin fold across the throat

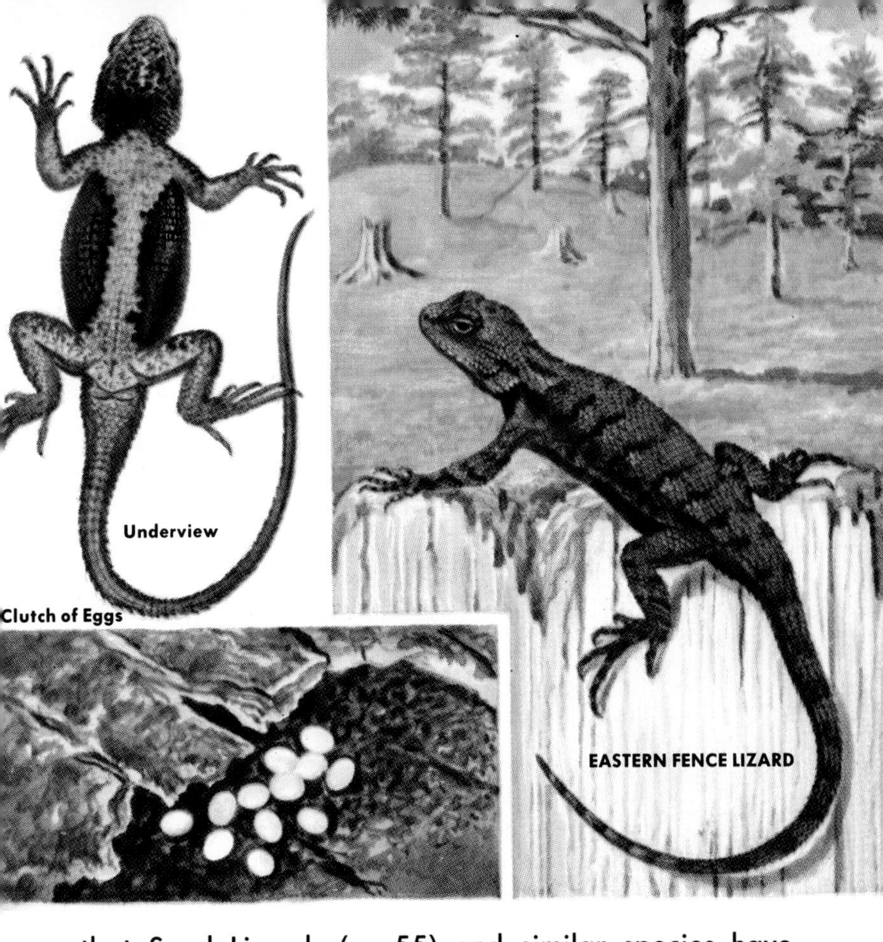

that Sand Lizards (p. 55) and similar species have. Some Spiny Lizards are blue or blue-patched on the underside; this is more pronounced in males. Detailed identification may be difficult. Spiny Lizards are good climbers; they are often found in trees, on boulders, among rocks. Their food is mainly small insects. Two species live in Florida, one elsewhere in eastern U.S. Others occur from Texas and Wyoming westward.

Spiny Lizards (16 species)

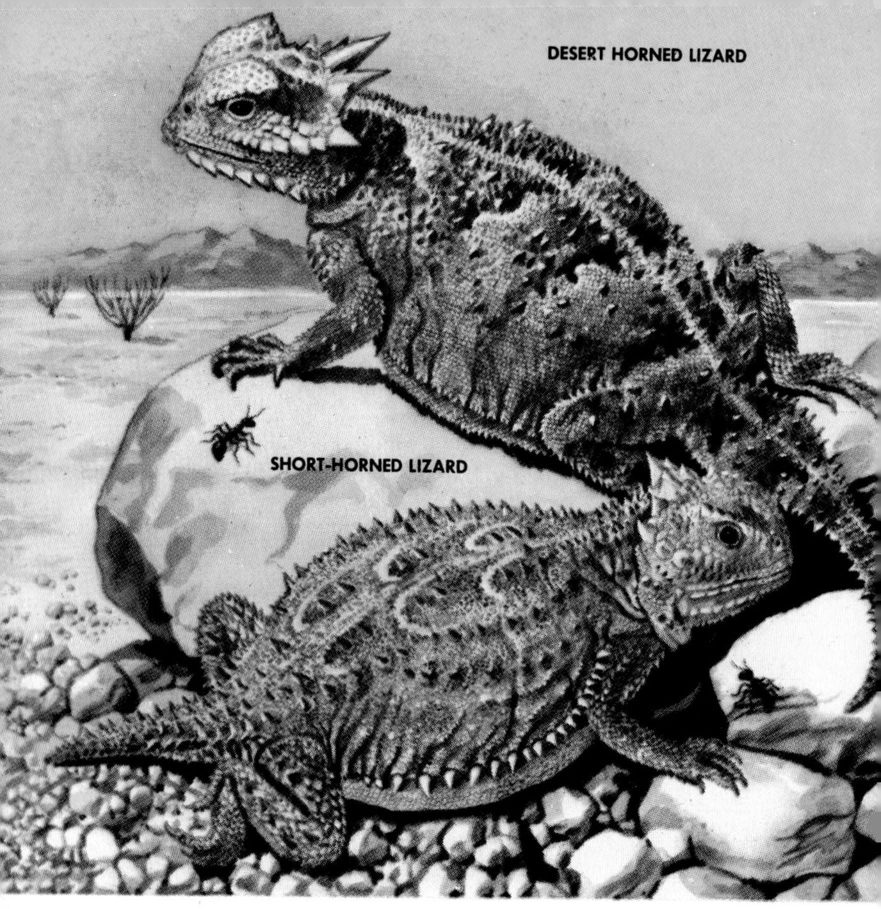

HORNED LIZARDS are unique. These odd, flattened creatures are found only in the West and in Mexico. The only other lizard like them is one in Australia. All have various-sized spines on the head which give the group its name. Seven species are found in dry, sandy areas, where they lie on rocks or half buried in the sand. When an insect appears, a quick snap of the lizard's tongue takes care of it. Some species lay 20 to 30 eggs; in others up to a dozen young are born alive. In one species eggs hatch in only a few hours; others take several weeks. These unusual liz-

ards may squirt a thin stream of blood from the corners of their eyes when frightened. Some puff up when angered; others flatten themselves out even more. Horned lizards hiss threateningly, jump at an intruder, lift themselves high on their legs, and tilt their body toward danger. All such behavior is a bluff, however; the animals rarely bite. Protected in most states, they should never be kept as pets. Ants are their favorite food, but they do not eat well in captivity.

NIGHT LIZARDS are mottled, medium-sized lizards. Both body and tail are about 3 in. long. They live in areas of granite, behind the loose-scaled flakes of rock or under fallen stalks of yucca. Note the vertical pupil in the eye and lack of eyelids. Horizontal rows of plates cross the belly. The three species are nocturnal, spending the day in sheltered cracks. Young (two or three at a time) are born alive. The food is beetles and other small insects.

SKINKS Some 15 species of Skinks are found in the United States; no other lizards have so wide a range. They are the only lizards that many people in the North have ever seen. All are small to moderate in size. The body length is not usually more than 5 in., the tail not much over 6 in. Most Skinks are smaller. Skinks can be recognized by their smooth, flat scales, which produce a glossy, silky appearance. Most have short legs, and in one (p. 64) the legs are degenerate, but most are swift runners. All burrow occasionally, for Skinks, in general, are ground lizards. Active during warm days, Skinks feed mainly on insects, spiders, worms, and perhaps small vertebrates. They hibernate all winter in the ground or under logs. The most common Skink mates during May. Eggs, 6 to 18, are laid about six weeks later. The mother spends the next six or seven weeks brooding her eggs till they hatch, something unusual for lizards. The young are only about an inch long.

Skinks can be roughly identified by the markings on their backs. Most common in the East are the five-lined Skinks, which have five light lines from head to tail. Lines are clearer in younger animals. In the West, four-lined Skinks are common. Other Skinks have eight lines, two lines, or no lines at all. The larger species have powerful, narrow jaws and inflict painful bites, the body often thrashing and twisting vigorously as the jaws maintain their grip.

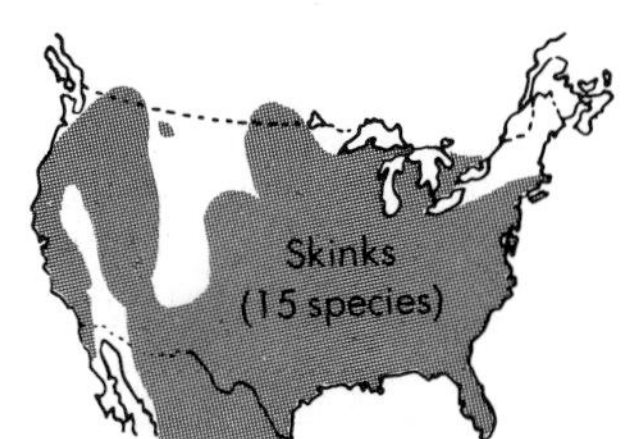

BROADHEAD SKINK
GILBERT'S SKINK
WESTERN SKINK
GREAT PLAINS SKINK

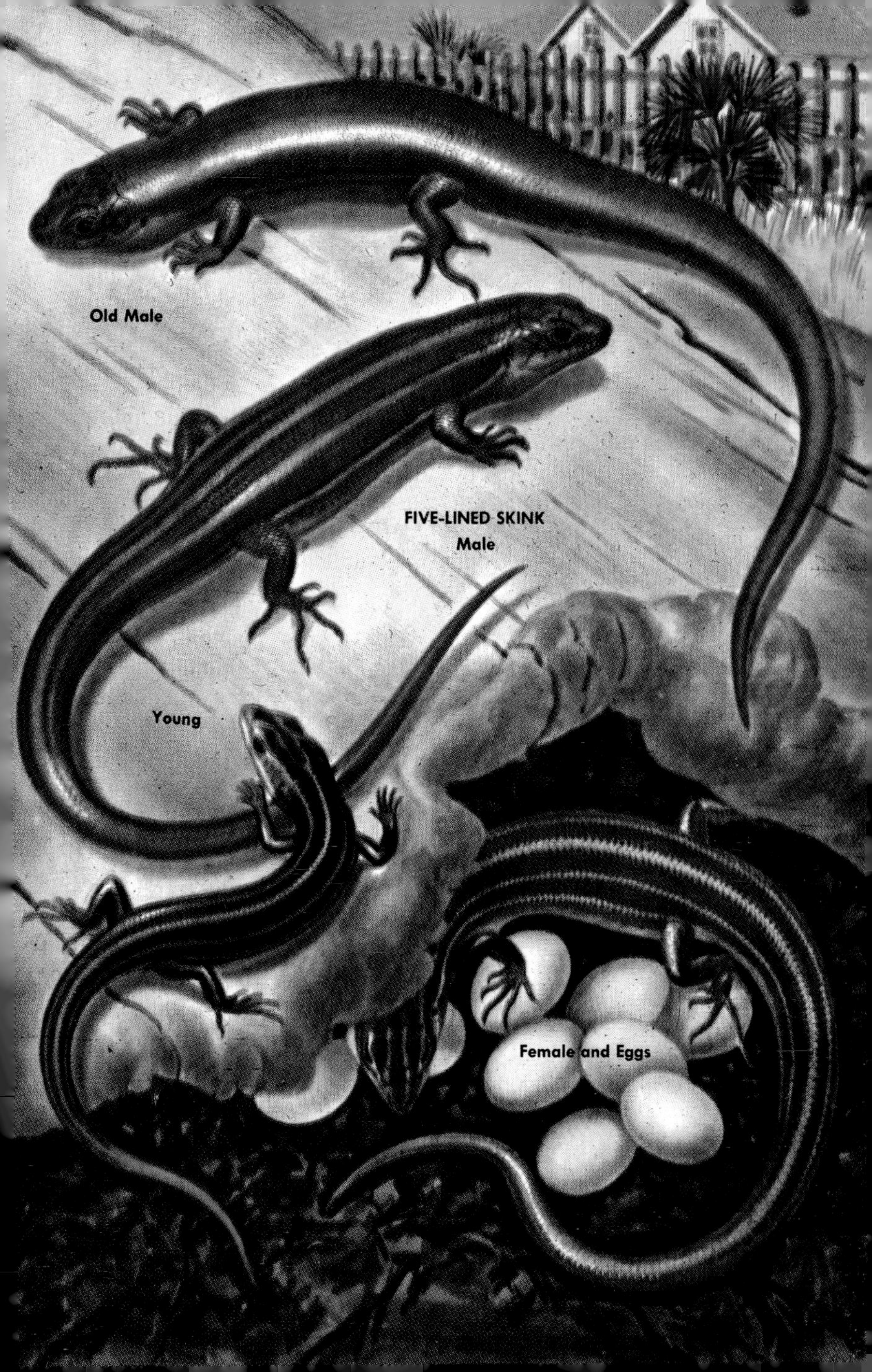
Old Male
FIVE-LINED SKINK
Male
Young
Female and Eggs

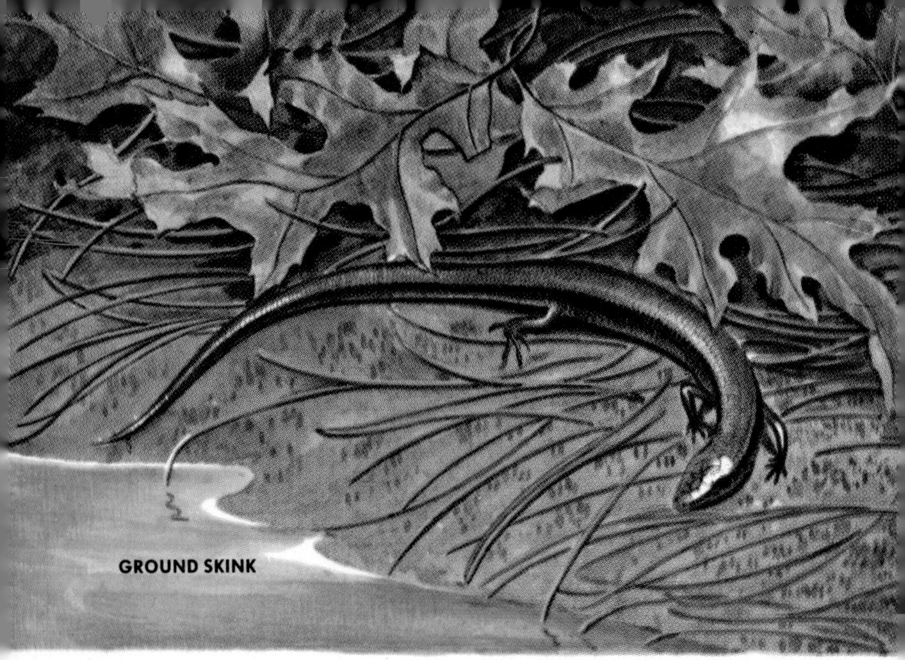

GROUND SKINK

GROUND and SAND SKINKS are similar to the other skinks just noted. The Ground Skink (2 in. long, with longer tail) has a transparent disk in the eyelid. Notable also are the smooth, flat scales and the broad, brown bands down the sides. The skink prefers wooded moist places; it lays its eggs in humus or rotted wood. It is an active lizard, most commonly found on the ground, often hiding under leaves. The Sand Skink is a burrowing lizard about 2 in. long, with legs small and degenerate, especially the forelegs. No other lizard has legs quite like it. It is found in pine woods, in dry or sandy soils.

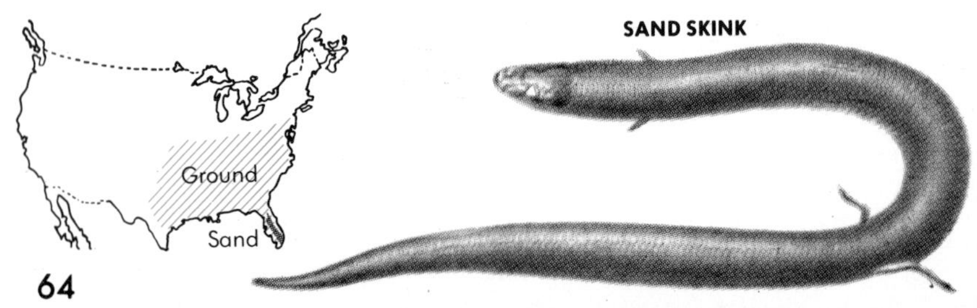

WHIPTAILS and RACERUNNERS give the experts trouble. The common and widespread Racerunner (one species) is the only species of the eastern half of the country. Its body length is about 3 in.; tail at least twice as long. These active lizards are found in dry, open areas, feeding during the day on insects, worms, and snails. Whiptails (14 species) are checkered or spotted. They occur west of the Great Plains.

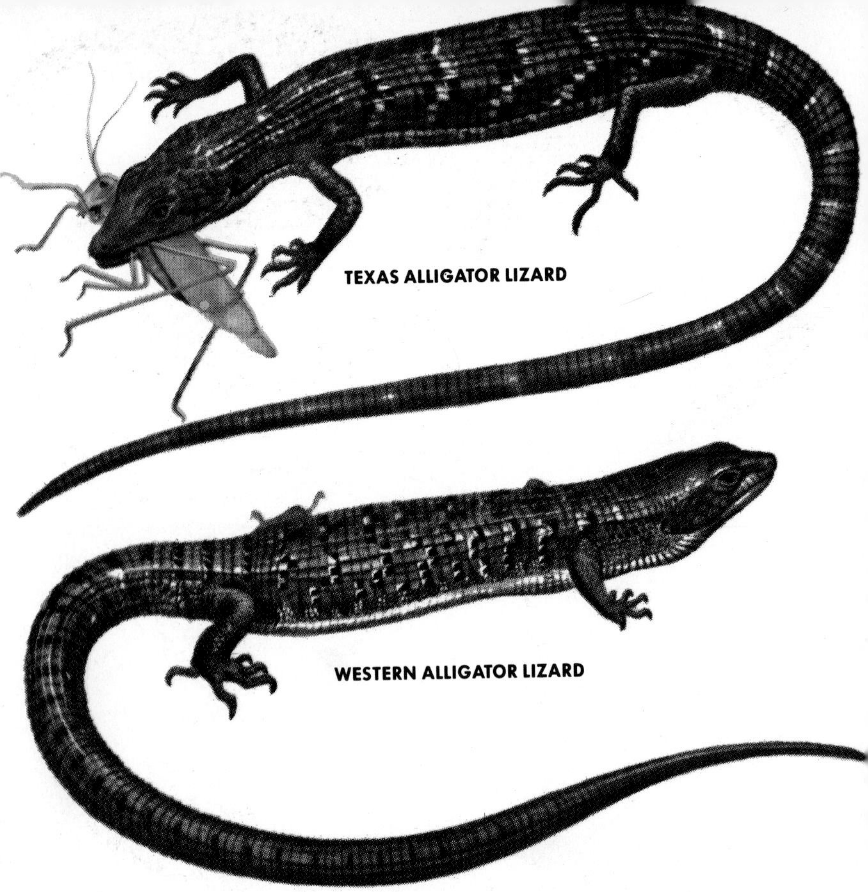

ALLIGATOR LIZARDS, named for their shape and heavy scales, are slow, dull-colored, solitary, with a banded or speckled back. Fairly large (to 8 in., tail twice as long if complete), they feed on insects and spiders and, in turn, are food for larger reptiles, mammals, and birds. A skin fold with tiny scales along sides of body allows for expansion of the abdomen with food or eggs. Males may bite painfully. Five species: four lay eggs; in one, young are born alive.

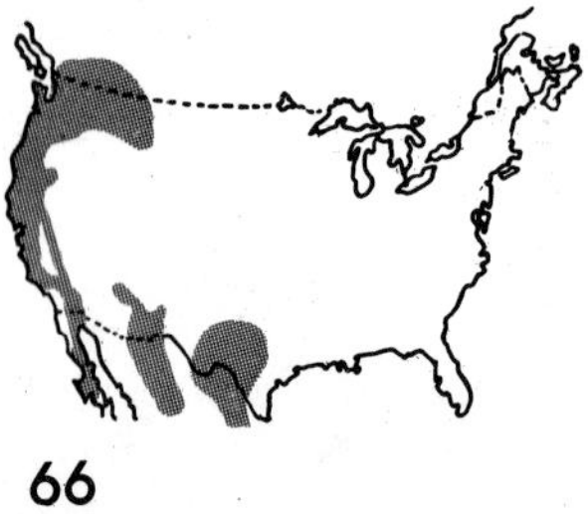

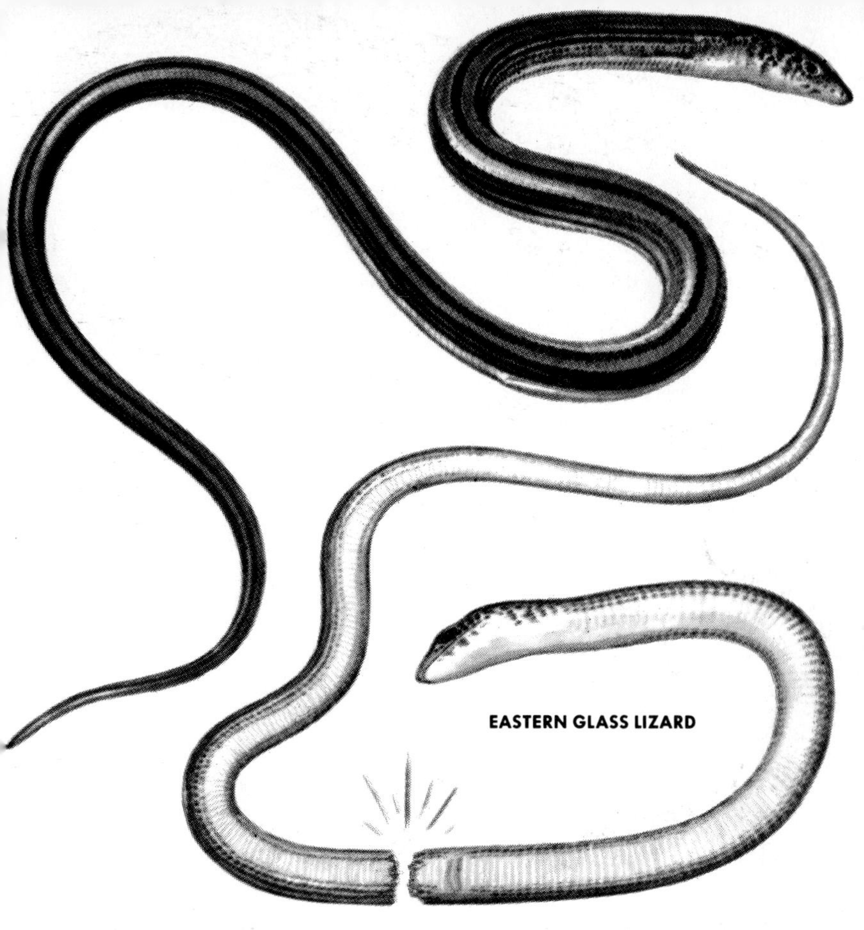

EASTERN GLASS LIZARD

GLASS-SNAKE LIZARDS are of three closely similar species—limbless, somewhat snakelike, 2 to 3 ft. long. Ear openings, eyelids, and many rows of belly scales proclaim them to be true lizards. The very long tail breaks off more easily than that of other lizards. It may break off when the animal is captured or roughly handled. The tail, of course, cannot rejoin the body, but a new, shorter tail grows in its place. These lizards feed on insects. They may bite when handled.

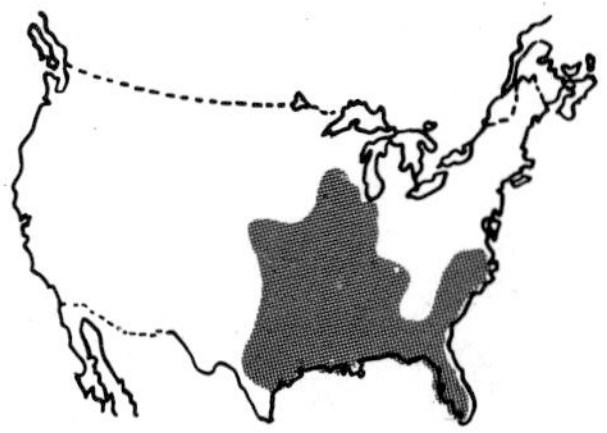

FLORIDA WORM LIZARD

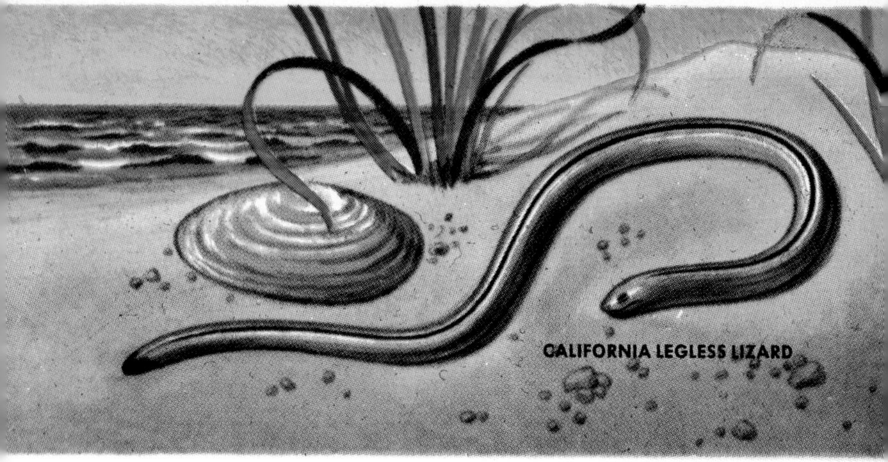

CALIFORNIA LEGLESS LIZARD

WORM and LEGLESS LIZARDS are two small burrowing species. The former (up to 10 in. long, only ¼ in. thick), found in sandy soil of pine woods, has distinct rings which make it look much like a large earthworm. Limbless, earless, and blind, it belongs to a tropical suborder of a rank equal to those of lizards and snakes. The California Legless Lizard, which is even smaller (6 in.), has small eyes but is earless and limbless. One form is silvery, the other black. All eat small insects.

GILA MONSTER, our only poisonous lizard, grows up to 2 ft. long. The poison, from glands in the lower jaw, is not injected, may not enter the wound when the lizard bites, but is deadly. Usually slow and clumsy, Gila Monsters can twist their heads, bite swiftly, and hang on strongly. It is illegal to capture or molest them. Gila Monsters live under rocks and in burrows by day. They feed on eggs, mice, and other lizards. The 6 to 12 eggs hatch in about a month.

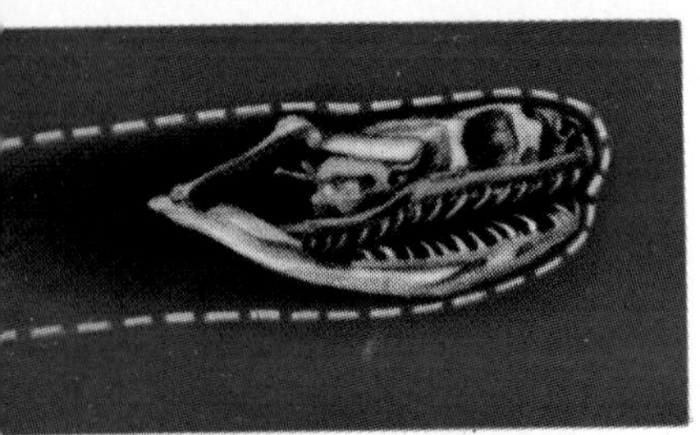
Skull of Non-poisonous Snake—
Eastern Racer

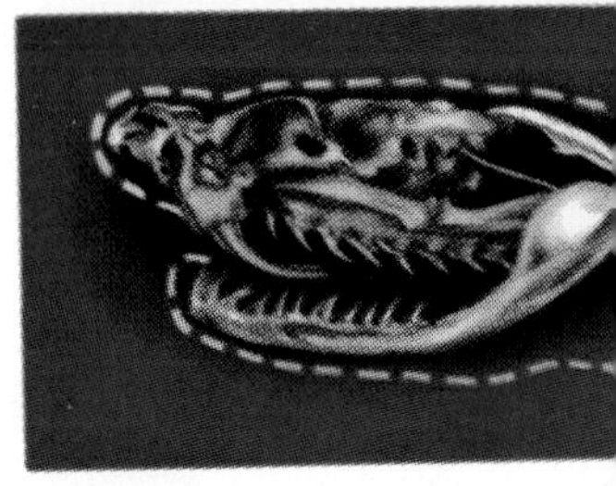
Skull of Poisonous Snake—
Cottonmouth

SNAKES are the best-known reptiles. Of some 120 species found in the U.S., 17 produce poison which can seriously harm man. In few places are poisonous snakes common, and death from snake bite is a rarity. All snakes except Blind Snakes have large scales across the belly. Besides lacking limbs, they also lack ear openings and eyelids that move. Each side of the snake's lower jaw moves separately, enabling it to swallow prey larger than its normal mouth size. Teeth are small and hooked. The larger fangs of poisonous species are grooved or hollow.

The snake's long, forked tongue is harmless. As it is protruded and flickered in the air, it collects odorous particles and brings them into the mouth and into contact with smell-sensitive organs there. These supplement the sensations the snake receives through its nostrils.

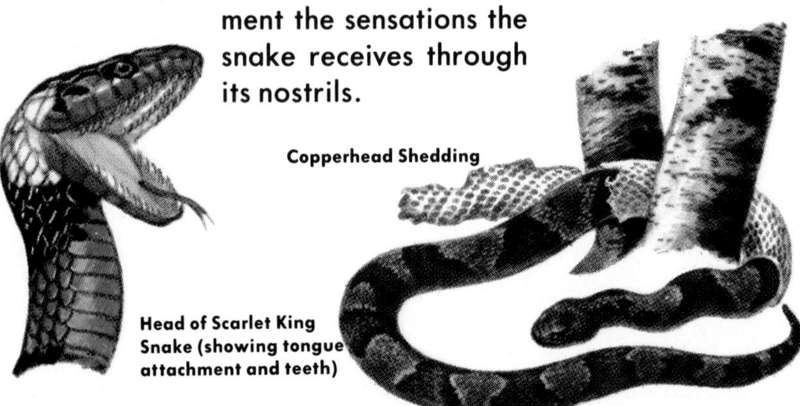
Copperhead Shedding

Head of Scarlet King Snake (showing tongue attachment and teeth)

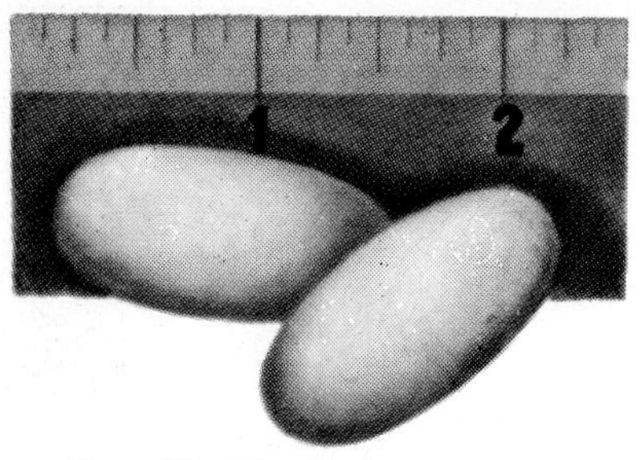

Eggs of Red King Snake

Copperhead at Birth

Snakes do not react to air-borne sound waves as we do; they respond more obviously to vibrations through the ground. The eyesight of snakes is fairly good, though their distance vision is poor. Some can see very well at night.

Snakes feed on live animals: insects, worms, frogs, mice, rats, and rabbits — mostly animals harmful to man. Some snakes lay eggs; the young of others are born alive. Snakes may have about a dozen young at a time and occasionally as many as 99. The mother gives them no care after birth; the young fend for themselves and grow rapidly. Most of them double their size in one year and are full grown in two or three years. In growing, snakes shed their skins at least once and often several times a year.

Although snakes are kept in captivity by some people, it is not recommended. Local, state, and federal restrictions limit possession of many species. They are nevertheless unusual, attractive — even beautiful. Not all eat well when caged. Live food is usually needed.

Shed Skin

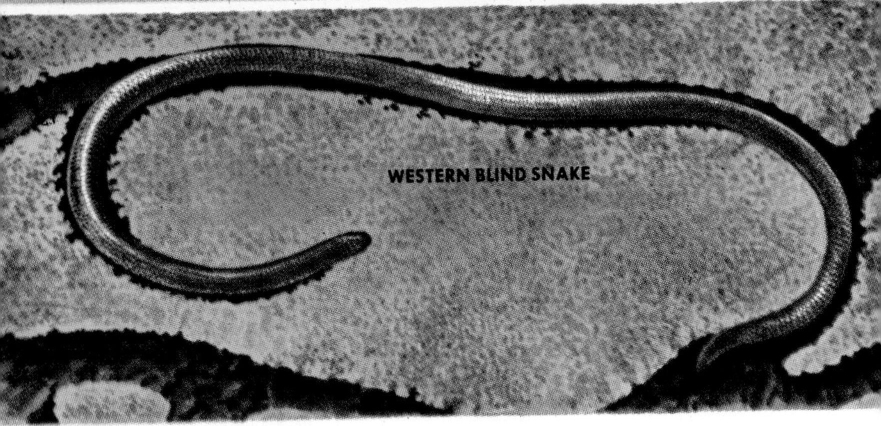

BLIND SNAKES, wormlike in color and size (8 to 12 in.), are truly blind. They may come to the surface at night. Most are found under stones or in digging. They eat worms and insect larvae. Captive specimens never bite. They burrow rapidly in moist sand or gravelly soil, remaining close to the moisture line. The two similar species are the only American snakes without large belly scales. Blind Snakes lay eggs. They are relatives of the Boas.

BOAS are not all large tropical snakes. Two species live north of Mexico. The Rosy Boa, an attractive, docile, small-scaled constrictor, lives in dry, rocky foothills. The grayish Rubber Boa, also heavy-bodied, has a short, blunt tail which it displays like a head while its real head is protected by the coils of its body. It grows up to 2 ft. long; the Rosy Boa is larger (3 ft.). Both bear live young, and are protected throughout their ranges.

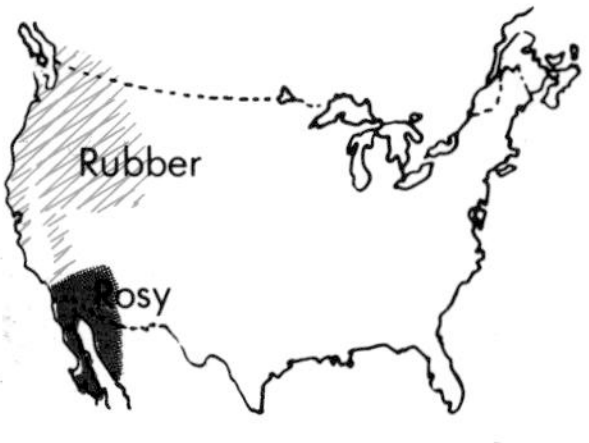

RAINBOW SNAKE is a handsome species. Stripes vary from orange to red. The underside is red with a double row of black spots. This snake of swampy regions often burrows and is not commonly seen. It is smaller (40 in.) than the closely related Mud Snake (p. 75) and like it has a sharp "spine" at the end of the tail. Little is known of its life history and feeding habits. The female Rainbow Snake lays 20 or more eggs, which hatch in about 60 days. Two races.

MUD SNAKE is the subject of many superstitions. The spike or stinger on the tail is erroneously said to be poisonous. This snake, also called Hoop Snake, is supposed to grasp tail in mouth and roll like a wheel. Such tales about the harmless, attractive, small-headed Mud Snake are untrue. This burrowing swamp snake feeds on fish and frogs, especially on Sirens and Amphiumas (pp. 139-140). Length, 4 to 6 ft.; lays 20 to 80 or more eggs. Two races.

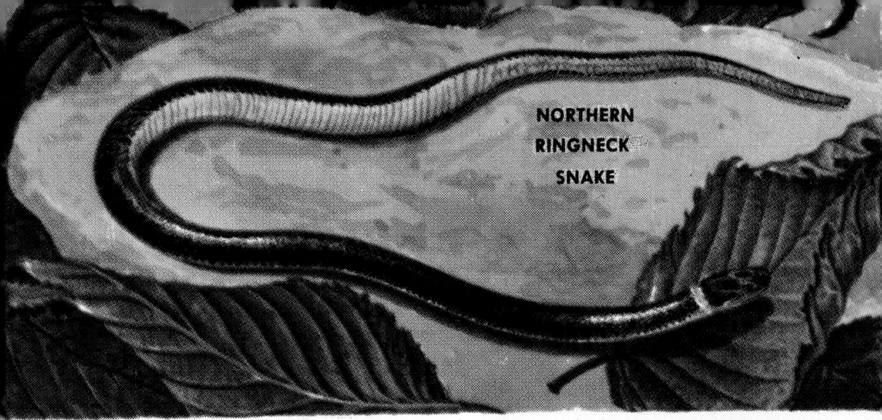

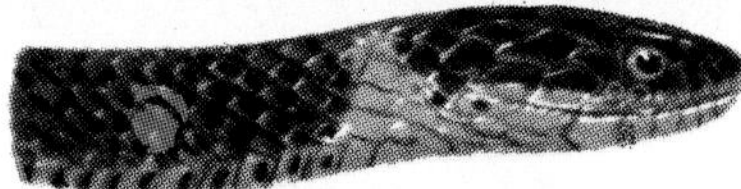

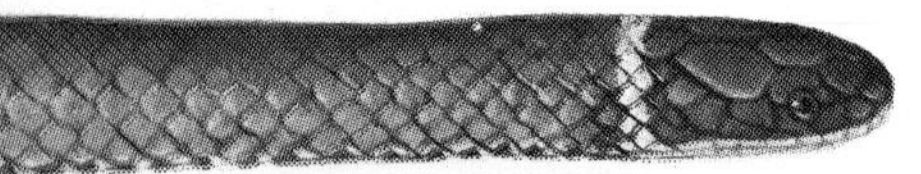

Eastern Group of Ringneck Snakes

Variations in Belly Markings

Northern Southern Prairie

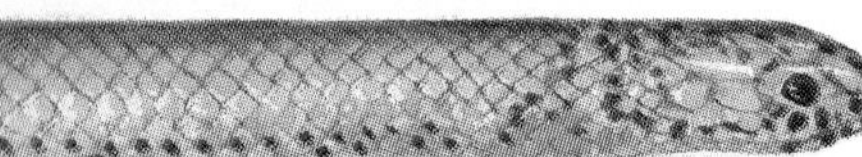

Western Group of Ringneck Snakes

RINGNECK SNAKE, one species with 12 subspecies in the U.S., is a small (12 to 18 in.), attractive snake living in moist woods under rocks or fallen logs, where it feeds on small insects and worms. It lays eggs hatching in about two months. Recognize this snake by its slate-gray color and usually the yellow-to-orange ring behind the head. The underside is yellow, orange, or red, sometimes spotted. It may secrete a smelly fluid when captured, but does not bite.

GREEN SNAKES, slender and harmless, live in greenery where they are hard to see. The smaller species (15 to 18 in.) with smooth scales prefers open grassy places. The other, which grows twice as long, has a rough appearance due to a ridge or keel on each scale. Often found in bushes and vines, this one feeds on insects. Eggs of both species hatch into dark young which gradually turn green. Green Snakes are docile, but as neither eats well, they languish in captivity.

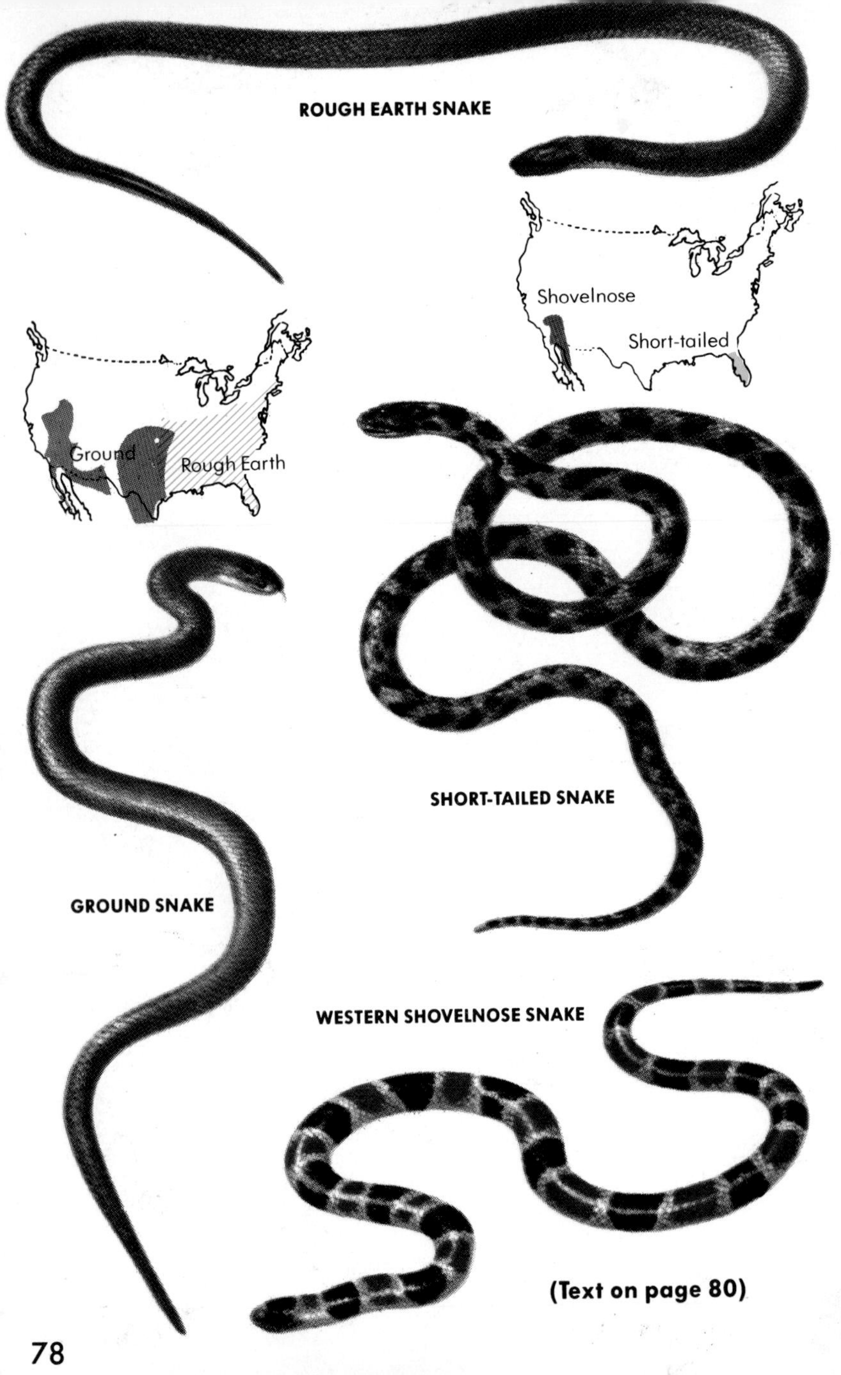

(Text on page 80)

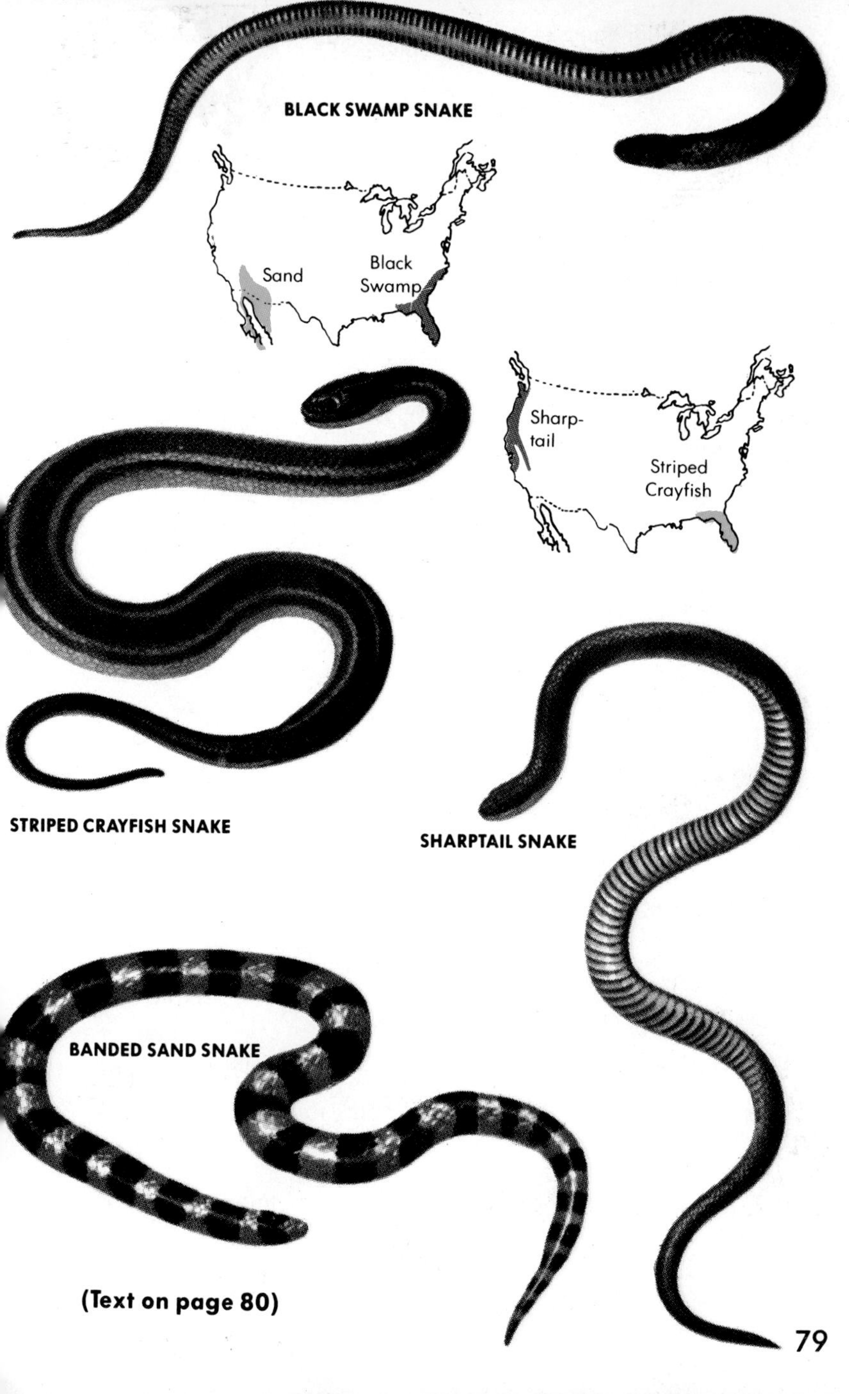

(Text on page 80)

SMALLER, LESS COMMON, HARMLESS SNAKES

(Illustrations on Pages 78 and 79)

EARTH SNAKES (10 to 12 in.) are two woodland species, one Rough, the other Smooth. Brownish or gray above, some with small black dots. Food: small insects and worms. Young are born alive.

SHORT-TAILED SNAKE (18 to 24 in.) is like a small, slender Eastern Milk Snake (p. 98). An aggressive, burrowing, upland snake, it kills small prey, often other snakes, by constriction. Tail is very short.

GROUND SNAKE (10 to 15 in.) is a small species of exceptionally variable color and pattern, similar but not related to Sharptail. Food: insects, spiders, etc.

SHOVELNOSE SNAKES (12 to 16 in.) are ground snakes (two species) slightly larger than Ground Snake and related to it. Snout projecting but flattened. A yellowish, egg-laying sand burrower.

BLACK SWAMP SNAKE (12 to 16 in.) is thick-bodied, red-bellied, swamp-loving. Black bar on each belly scale. Young born alive. Food: probably fish, frogs.

STRIPED CRAYFISH SNAKE (18 to 24 in.) is aquatic, living in holes and tunnels along ditches and in swamps. Food: mainly crayfish and frogs. Young are born alive.

SHARPTAIL SNAKE (12 to 16 in.) is somewhat stout. Little is known of its habits. Note the light yellow stripe on sides, black bands on yellow belly scales.

BANDED SAND SNAKE (10 to 14 in.) is a burrower in desert sands. Crawls just below the surface, aided by a broad, heavy snout. Yellow to red, with dark bands almost encircling body. Scales small and shiny. Life history largely unknown. Said to eat ant larvae.

WESTERN HOGNOSE SNAKE

EASTERN HOGNOSE SNAKE

HOGNOSE SNAKES are unique and amusing. When molested, they hiss, spread the head, and strike, as though to appear dangerous, but they never bite. If threats fail, they roll over and play dead. Because of their ferocious puffing, these harmless snakes are sometimes called Puff Adders. The hard, turned-up nose helps in burrowing after their favorite food, toads, which often make up their entire diet. A Hognose Snake lays about two dozen eggs in summer. These gentle snakes do well in captivity if fed toads or lizards. Eastern Hognose Snake, 2 to 3 ft. long, is the largest of three similar species, all heavily built. Southern species, like Western, has sharply turned-up nose.

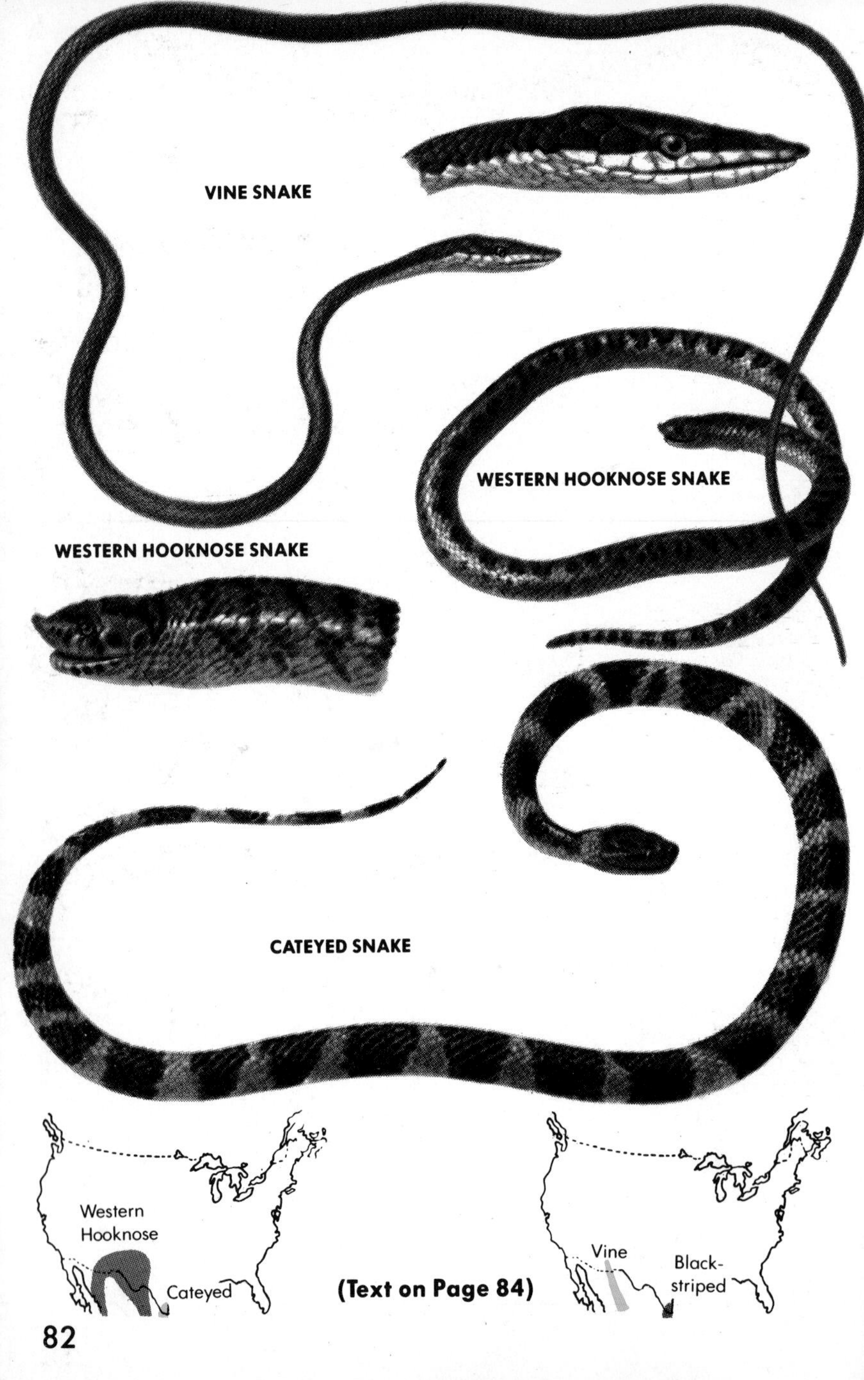

(Text on Page 84)

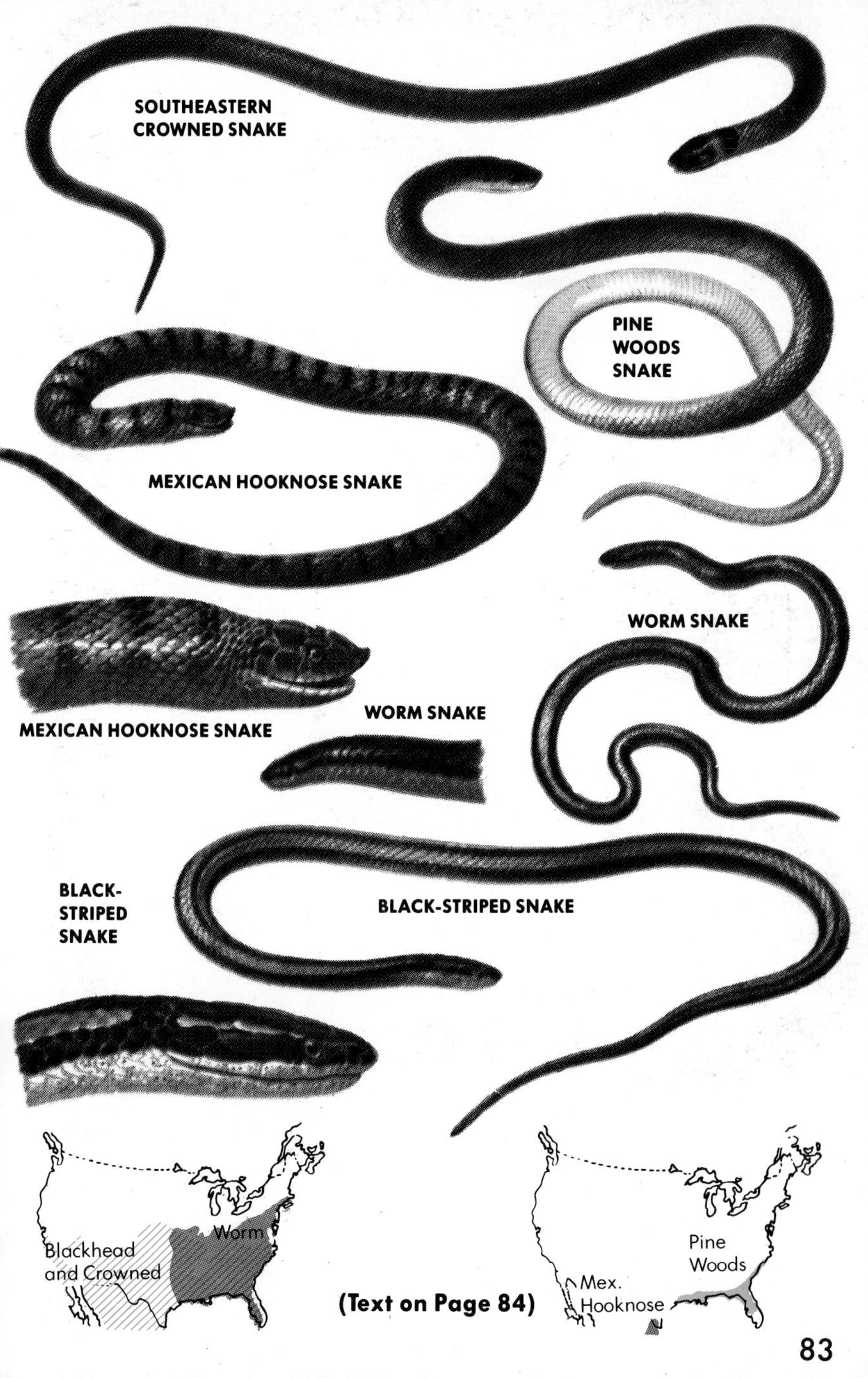

(Text on Page 84)

OTHER LESS COMMON SNAKES

(Illustrations on Pages 82-83)

VINE SNAKE* (to 4 ft. or more) is a bush-dweller of semiarid regions, very slender. Longer, narrower head than other American snakes. Reddish-brown; white line down belly. Food: lizards.

WESTERN HOOKNOSE SNAKE is a blotched, egg-laying burrower resembling a miniature Hognose Snake (10 to 12 in.) but is not kin.

CATEYED SNAKE* (to 30 in.), wide-headed, slender, is an egg-layer. It feeds on both invertebrates and small vertebrates. Often found in trees and bushes.

BLACKHEAD and CROWNED SNAKES* (12 to 14 in.) are a large group of secretive or burrowing, egg-laying species. All but one have a black head cap.

PINE WOODS SNAKE (12 to 16 in.) has a yellow upper lip. Back reddish-brown, belly yellow. An egg-layer of swamps, found under logs and debris. Food: frogs, toads, insects.

MEXICAN HOOKNOSE SNAKE (10 to 12 in.), related to the Western, is subtropical, with a larger shovel-snout than the Western but with similar habits. Ashy gray with gray and black cross bands.

WORM SNAKE (10 to 13 in.) is a burrower, rarely seen. Shiny, smooth scales. An egg-layer; feeds on earthworms. Found in woods.

BLACK-STRIPED SNAKE* (16 to 20 in.) is a night snake. Eats frogs, toads, lizards. An egg-layer and ground-dweller. Rare; commoner in tropical America.

*Species with weak venom and small, fixed, grooved fangs in rear of upper jaw.

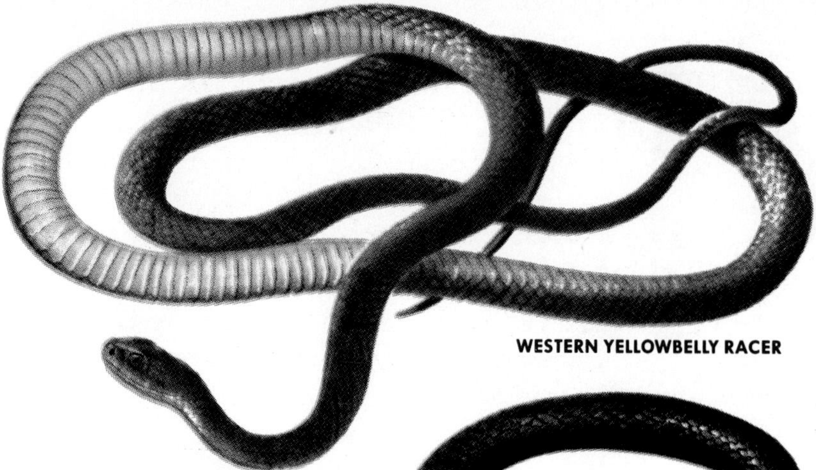

WESTERN YELLOWBELLY RACER

NORTHERN BLACK RACER

RACER (one species, with ten subspecies) is aggressive and graceful. In the East, they average 4 ft., are smooth, blue-black, with white chin and throat. Western form is smaller, greenish or yellowish brown, with belly and chin lighter. All Racers are very active, at home in bushes and trees. Their food is small mammals, birds, insects, frogs, lizards, and other snakes.

YOUNG RACER

EASTERN COACHWHIP SNAKE

COACHWHIP and WHIPSNAKES are closely related to the Racer, but longer. The former (one species, seven races) is a variable brown (some are red or pinkish), darker at the head in some, becoming lighter toward the tail. Coachwhips are the most widely distributed and the largest of the group; some over 7 ft. long have been reported. Whipsnakes (three species) are usually 4 to 5 ft. long. These are typically striped with yellow on the sides against a dark back; the belly is usually lighter. Several of these are desert forms, but all

DESERT STRIPED WHIPSNAKE

are active during the day. All are alert and fast. They feed on mice, lizards, and small snakes, moving rapidly over sand or through brush after their prey. They do not kill by constriction, but chew when they bite. Too swift to be caught easily, when they are caught these snakes will strike and will thrash their thin, long tails. They are always aggressive and nervous. Eight to 12 eggs are laid in early summer, each about 1 by 1½ in.

PATCHNOSE SNAKES (three species) are as fast and active as the related Racers. On the move day or night, they like any terrain. The blunt shield over the nose plus the yellow and brown stripes down the snake's back give positive identification. Adults are about 3 ft. long. Lizards and other small desert life are eaten. Females lay eggs. The nose may help in burrowing in sand. Rear teeth are enlarged but venom is weak.

SADDLED LEAFNOSE SNAKE

LEAFNOSE SNAKES, two small relatives of the Patch-noses, have an even more exaggerated nosepiece. Once considered very rare, they are fairly common in the deserts at night. The two species are marked by dark blotches. Both are moderately stout, 12 to 15 in. long. They are pugnacious, coiling and striking when caught, but are harmless. They are egg-layers, and are reported to feed on desert lizards or lizard eggs.

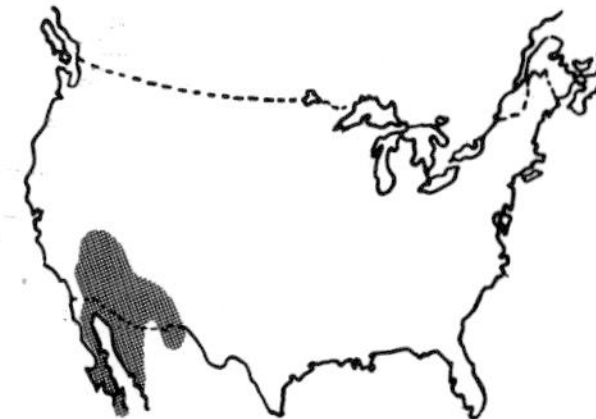

GRAY RAT SNAKE

RAT SNAKES include five large species, common and widely distributed in the East and Middle West. Colors differ from species to species, making identification easier. All are fast, active snakes. When caught they may bite freely and excrete a foul-smelling fluid from glands at base of the tail. They are not consistently aggressive, but bite unpredictably. All are constrictors. Members of this group have been known by diverse common names which are often misleading. Names used here are truer to the snakes. Black Rat Snake, also known as Pilot Black Snake, may be mistaken for the Black Racer (p.

YELLOW RAT SNAKE

BLACK RAT SNAKE (adult)

85). Black Rat Snake has some scales tipped with white —remains of a pattern of blotches seen more clearly in young of all members of this group. The scales are slightly keeled; those of the Black Racer are not. Gray Rat Snake, a more southern form, has blotches of gray or brown against a lighter background. Yellow Rat Snake (Striped Chicken Snake) averages 4 to 5 ft. long, sometimes reaches 7 ft. It is dull or olive yellow with four black lines down its back. Often found around barns or stables, it is looking for rats more often than for fowl. The Black, Gray, and Yellow Rat Snakes are all subspecies of one species, here called the Common Rat Snake.

BLACK RAT SNAKE (young)

CORN SNAKE

CORN and FOX SNAKES are more colorful members of the rat snake group. Corn Snake (often found in corn fields) is also known as Red Rat Snake because of the reddish-brown or crimson blotches against the lighter background. A very similar western form lacks the red color. Corn Snake does not grow as large as Yellow Rat Snake but it exhibits (as do the others) a pattern of hissing and vibrating its sharp tail when cornered or molested. Fox Snake (two subspecies) is like the other rat snakes but is somewhat heavier. It averages 3 to 4 ft.

FOX SNAKE

long, with brownish blotches against a straw-yellow base color. Found in woods and open country, it climbs less than other rat snakes. All rat snakes lay eggs, often in rotted logs or stumps. The young, blotched in color, may differ much from adults. Corn Snake and Fox Snake feed and reproduce better in captivity than other members of the group. Both feed almost exclusively on mice. Similar to Pine Snake, but head less conical, flatter.

INDIGO SNAKE may reach over 8 ft. Related South American forms are even larger. Two subspecies in U.S. Eastern form is a heavy, handsome, shiny, midnight-blue, fast racer, feeding on small mammals and other snakes. It is often found in burrows of gophers or rabbits. The Texan subspecies is brownish anteriorly, splotched. Both are rigidly protected, having become rare, especially in the East. Bite unpredictably, without striking, and chew tenaciously.

GLOSSY SNAKE is related and somewhat similar to the Bullsnake (p. 96). It has smooth scales; those of the Bullsnake are keeled. These smooth, shiny scales are responsible for the Glossy Snake's common name. Blotched, spotted, and gray-brown, these snakes are slender, with narrow heads. They are constrictors, feeding on lizards, rodents, and other small animals. They lay eggs and are nocturnal. Adults average 30 to 36 in. long. Seven races.

NORTHERN PINE SNAKE

Pine Snake Eggs

BULLSNAKES and their kin are found from coast to coast. These large, heavy snakes average 5 ft. long and grow up to 7 ft. They are the most common constrictors, widely known as destroyers of rodents. Bullsnakes have a large, vertical nose plate, adapted for burrowing. All hiss very loudly when angered and will strike to defend themselves, often vibrating the tail tip in warning. Most become docile in captivity, but others remain nervous. The Pine Snake is an eastern form of the Bullsnake, named for its favorite habitat—southern pine woods. It is relatively light-colored with large black patches on the back. Food is small rabbits, squirrels, rats, and mice.

NORTHERN PINE SNAKE

Farther west, the Bullsnake is more common. It is more yellowish than the Pine Snake and has a larger number of dark blotches. It often enters burrows to feed on pocket gophers and ground squirrels. The Pacific Coast form, known as Gopher Snake, is similar to the Bullsnake but smaller and with more blotches. All snakes in this group are protected against wanton killing. There is no doubt of their value as one agent in rodent control.

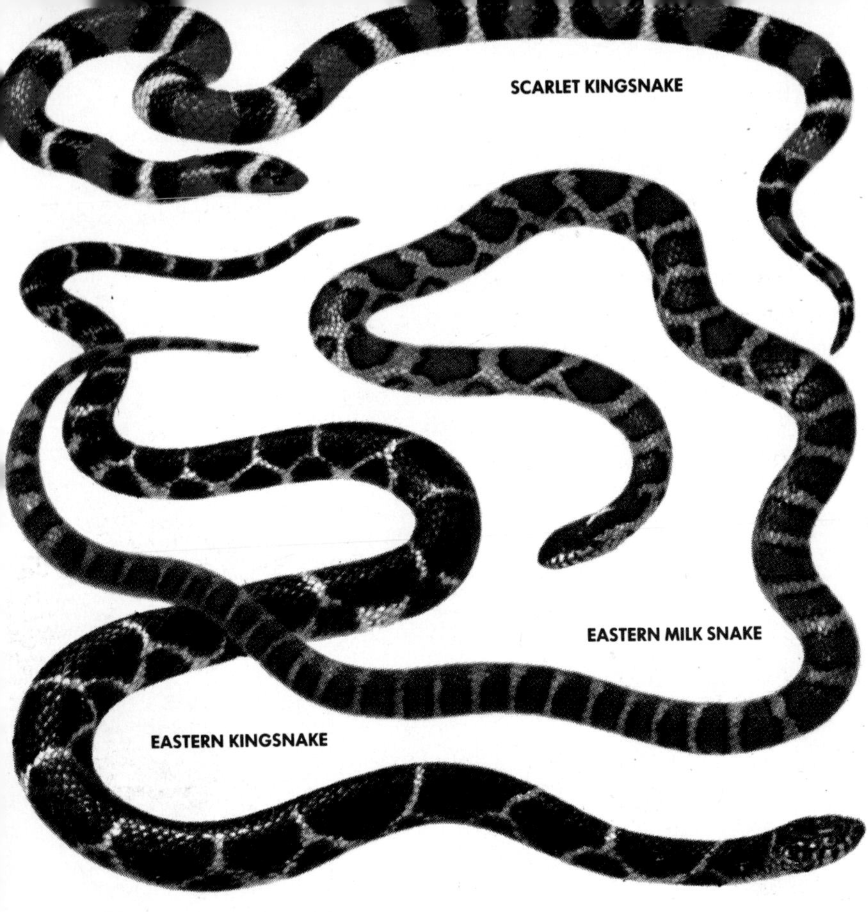

SCARLET KINGSNAKE

EASTERN MILK SNAKE

EASTERN KINGSNAKE

KINGSNAKES and MILK SNAKES are a group of medium-sized snakes of normal proportions. They include six species, ranging from southern Canada through much of the U.S. All are constrictors and some are at least partially immune to the poison of our venomous snakes. Kingsnakes feed on other snakes, but also eat many kinds of rodents. Milk Snake (30 in.) of central and eastern U.S. has red splotches or rings

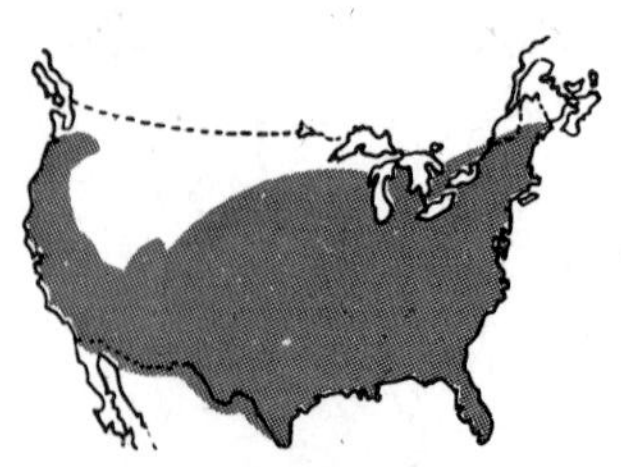

bordered with black; belly is pale, with black patches. It feeds mainly on rodents and not, as fables tell, by milking cows. Eastern Kingsnake is shiny black with bands of yellow crossing in a chainlike pattern. It is larger (3½ to 4 ft.) than Milk Snake and more common in open country. The small Scarlet Kingsnake (18 in.), like some races of the Milk Snake, may be confused with Coral Snakes (p. 108), but note that each yellow band is bordered by black. California Kingsnake (3 ft.) is black with white bands; some have a white midline. Speckled form has a light dot on most scales.

SPECKLED KINGSNAKE

CALIFORNIA KINGSNAKE

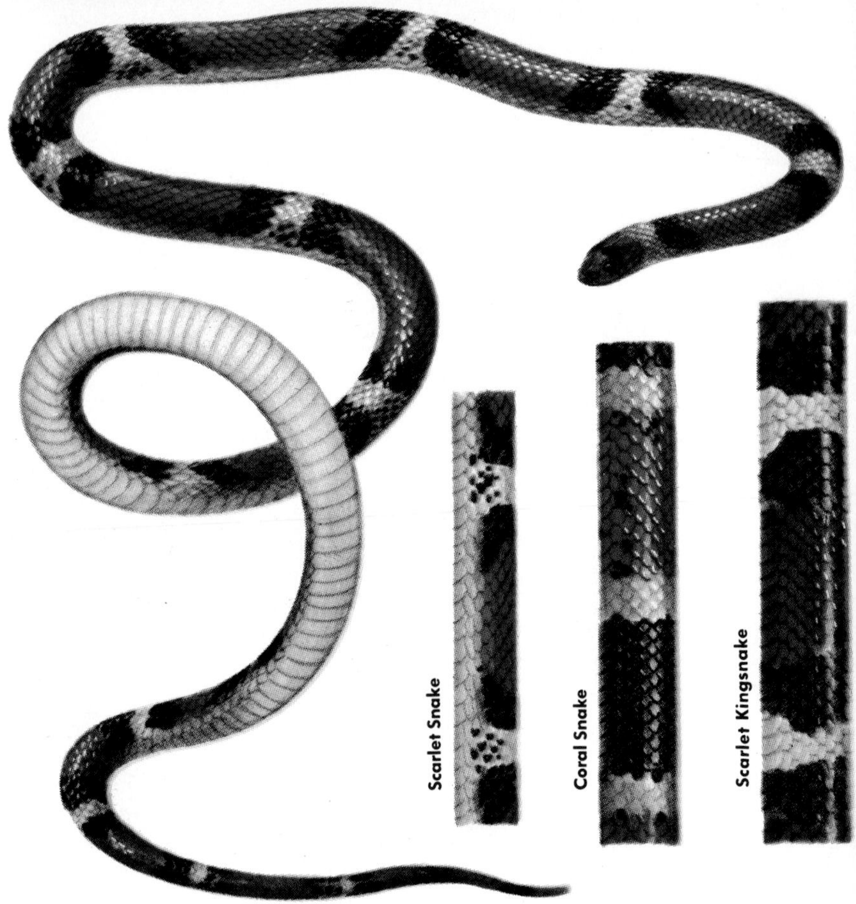

Scarlet Snake

Coral Snake

Scarlet Kingsnake

SCARLET SNAKE, unfortunately rare, is attractive and docile. It is a burrower, 16 to 24 in., occasionally found under rotting logs or on the ground at night. The food appears to be small lizards and mice, killed by constriction. Eggs are laid. The markings are similar in pattern to those of the Scarlet Kingsnake—yellow bands bordered by narrow black bands. But the bands do not encircle the belly. Coral Snake (p. 108) has black bands bordered by yellow. Three races.

LONGNOSE SNAKE (2 to 3 ft.) does not have a long nose. A burrower, it is aided by its small, narrow head. Most have been caught at night. It feeds on lizards, snakes, and small mammals killed by constriction. Dark patches on the back are broken by bands of red, white, or yellow. Generally speckled, the color is variable; belly lighter with a few dark spots. Longnose Snake is the only harmless snake with a single row of scales under the tail. Two races.

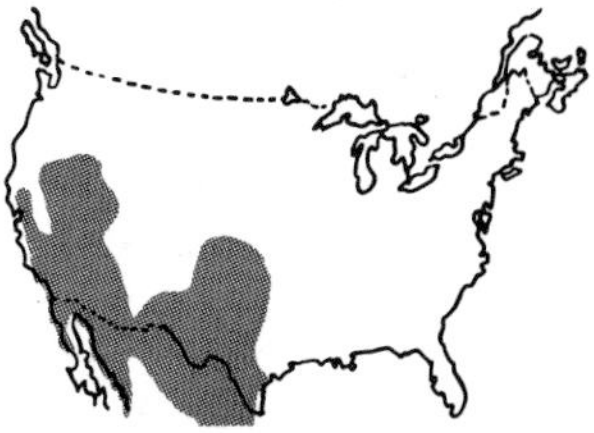

WATER SNAKES are chiefly eastern species (11). They show little external adaptation to water life but are actually fine surface and underwater swimmers. Most seek water when molested and there find their food, mainly crayfish, fish, and frogs. All are non-poisonous, but some are heavily built, with short, narrow tails, and are easily confused with the venomous southern Cottonmouth. Most Water Snakes are vicious, and, when caught, exude a repulsive-smelling fluid, mixed with feces, from glands at the base of the tail. The rear of the Northern Water Snake (30 in.) is cross-banded, with

reddish-brown. Toward the head these bands become large blotches. Diamondback Water Snake is larger (3½ to 5 ft.) and darker. Its dark blotches are reduced to diamonds over the backbone. Redbelly Water Snake (3 to 5½ ft.) is dark above, with a yellow or reddish belly. Green Water Snake (3 to 5½ ft.) is a dull olive green with a vague, barred pattern. Color and pattern are clearest in young Water Snakes. As many as 99 young are born alive.

PLAINS GARTER SNAKE

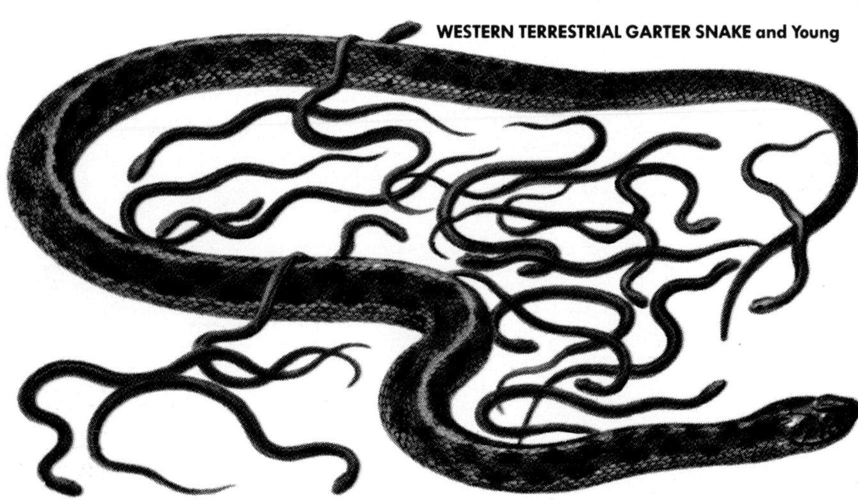
WESTERN TERRESTRIAL GARTER SNAKE and Young

GARTER SNAKES and their kin are perhaps more common and better known than any other snake. These 13 small (18 to 44 in.), striped species with keeled scales, related to Water Snakes, have similar habits. Like Water Snakes they eject an unpleasant fluid from vent glands when captured. Garter Snakes feed on frogs, toads, earthworms. Young are born alive in summer–20 or more at a time. Most Garter Snakes, fairly docile, do well in captivity.

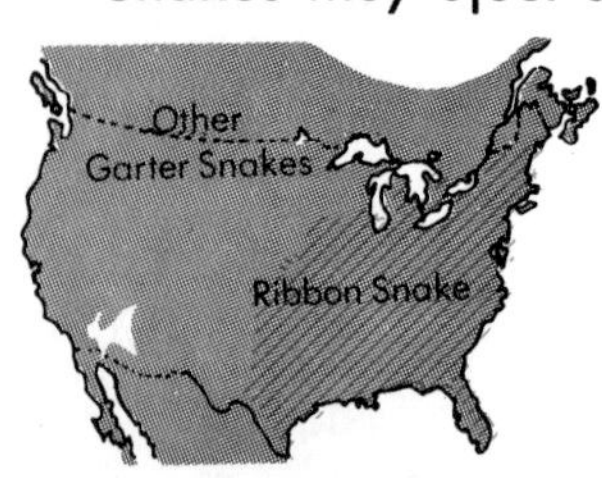

Common Garter Snake, more aggressive than others, is marked by three yellowish stripes; the dark area between is spotted. The center stripe of Plains Garter Snake is often a rich orange; the belly is darker than in Common Garter Snake. Several western species have the central stripe brighter than the side ones, and in others the side stripes are brighter than the median one. Ribbon Snakes (two species) are thinner, smaller, with yellow or red stripes against brown scales. Tail is nearly a third of body length.

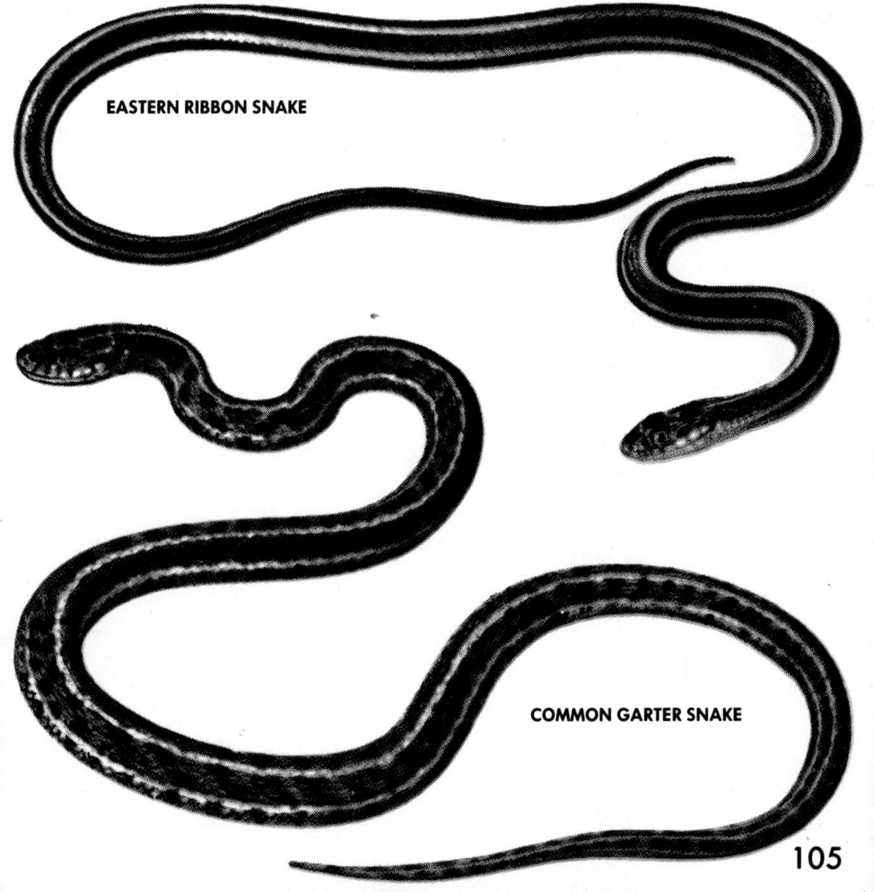

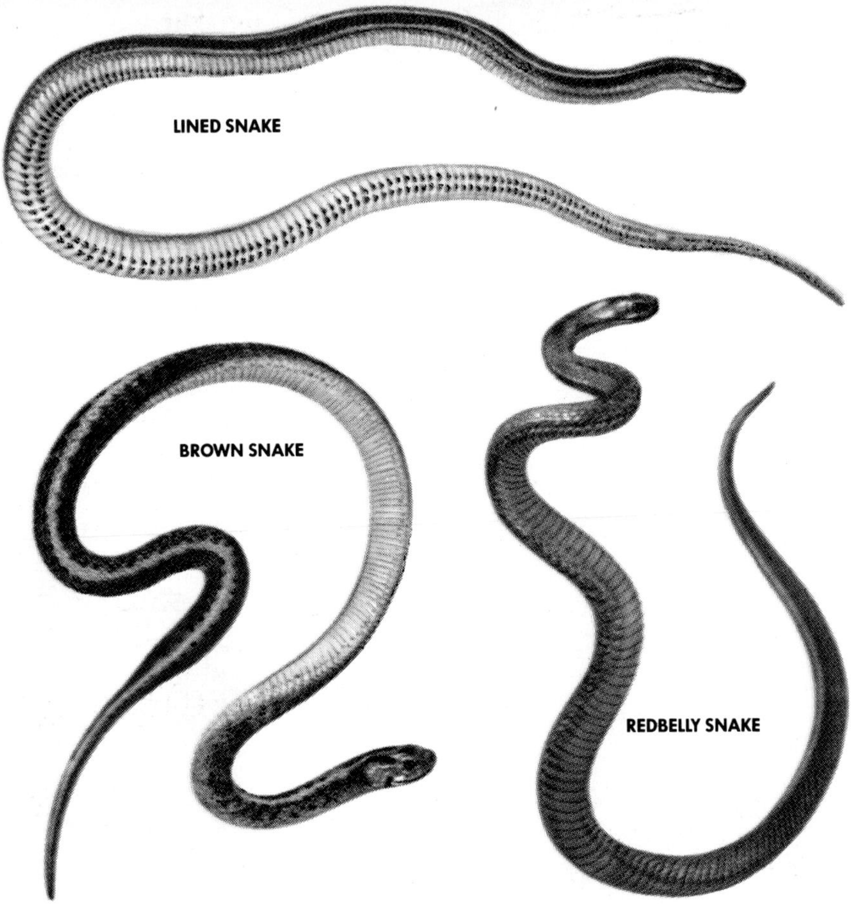

LINED SNAKE

BROWN SNAKE

REDBELLY SNAKE

SMALL STRIPED SNAKES are common but inconspicuous. Lined Snake (12 to 20 in.) is a miniature Garter Snake with a yellow stripe down its back, black dots on the belly. Brown Snake (10 to 16 in.) is a brownish, secretive, burrowing species, common even near cities. The belly is yellow to pink, with black dots at sides. Redbelly Snake (10 to 14 in.) is similar, but with red belly and yellow spots or collar at back of head.

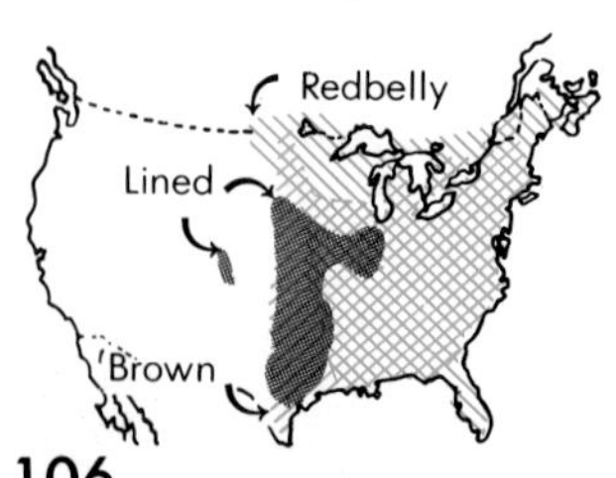

NIGHT SNAKE

LYRE SNAKE

NIGHT SNAKE and LYRE SNAKE are mildly poisonous. The former (15 in.) has enlarged teeth in the rear of its jaws, not true fangs. When it bites lizards, its poisonous saliva kills slowly. Lyre Snake (3 ft.) is a rear-fanged poisonous snake with grooved fangs. Its poison is not serious to man. The relatively large head and thin neck are characteristic of this snake. Lyre Snakes typically frequent rocky areas and feed on lizards.

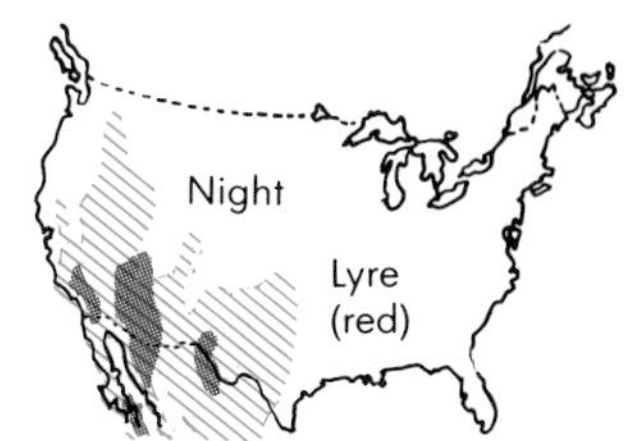

CORAL SNAKES, related to Cobras, are highly poisonous. Our two species have red, yellow, and black rings, the latter bordered by yellow. Eastern Coral Snake (30 to 39 in.) has black, yellow, and black from nose to back of head. Secretive and burrowing, it feeds mainly on lizards and other snakes. Western Coral Snake, smaller (18 in.), of limited range but similar habits, has black, yellow, and red successively on head and neck. Paralytic venom.

Western Eastern

COPPERHEAD and COTTONMOUTH are two poisonous species (one genus) related to rattlers. A pit between eye and nostril, sensitive to heat, helps them find and strike at warm-blooded prey. Copperheads (five races; 30 to 50 in.) are upland snakes, with coppery head and "hour glass" body patches. Cottonmouths (three races; 40 to 58 in.), larger, heavier, and more vicious, are swamp snakes feeding on fish and frogs. They are dark, not strongly marked.

Copperhead
Cottonmouth

Cross Section of Rattle

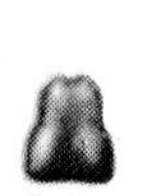
Button

Young

Older

Adult

Old Adult

RATTLESNAKES (15 species) are solely American. There are two major groups (genera): the Plated rattlesnakes, with large scales on top of the head (two species, see p. 111), and the Mailed rattlesnakes, with numerous small scales on top of the head. The Mailed rattlers (pp. 112-113) include two species in the East, 11 in the West. Timber Rattler (3½ to 6 ft.) is a woodland species, yellowish with dark, V-shaped bands and dark tail. Eastern Diamondback averages 5 ft. (record nearly 9 ft.). Westward is Western Rattler (2½ to 5 ft.), typically greenish yellow with darker blotches. Western Diamondback (4½ to 7½ ft.), of warm, dry areas, is brown-blotched. Red Diamond Rattler, similar to Western, is reddish. Strangest is Sidewinder (18 to 30 in.), with an erect scale over the eyes, and a rapid, sidewise motion over sand. The rattle gains a segment each time the snake sheds. Food: rabbits, gophers, rats, mice, other small mammals. Young born alive; litters of 12 are common. Venom deadly even at birth.

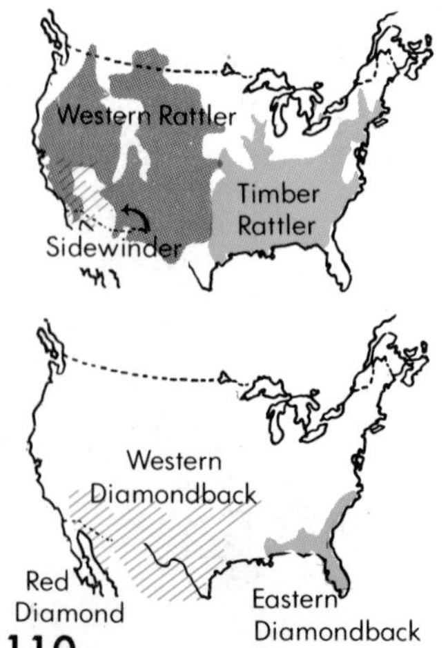

PIGMY RATTLER

MASSASAUGA

PIGMY RATTLER

TIMBER RATTLER

MASSASAUGA and PIGMY RATTLER, very similar to their larger relatives, have larger scales on the top of the head. They are small, hence relatively less dangerous. The Massasauga (three races; 2 to 3½ ft.), a swamp Rattler, does not strike unless much annoyed. The southern Pigmy Rattler (three races) is smaller (18 to 24 in.) and prefers upland terrain. Not mild-tempered, it rattles and strikes when approached. The rattle can scarcely be heard.

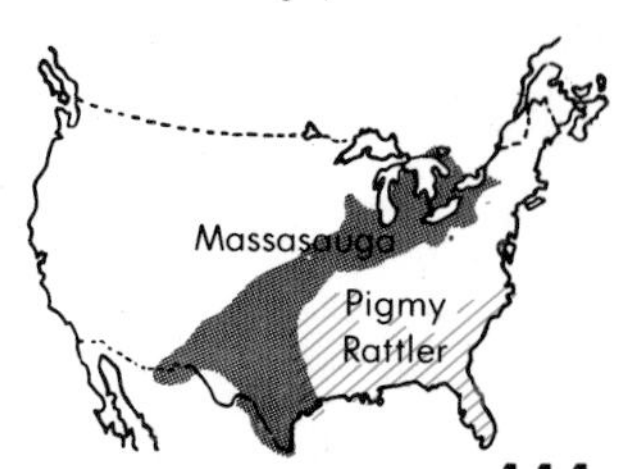

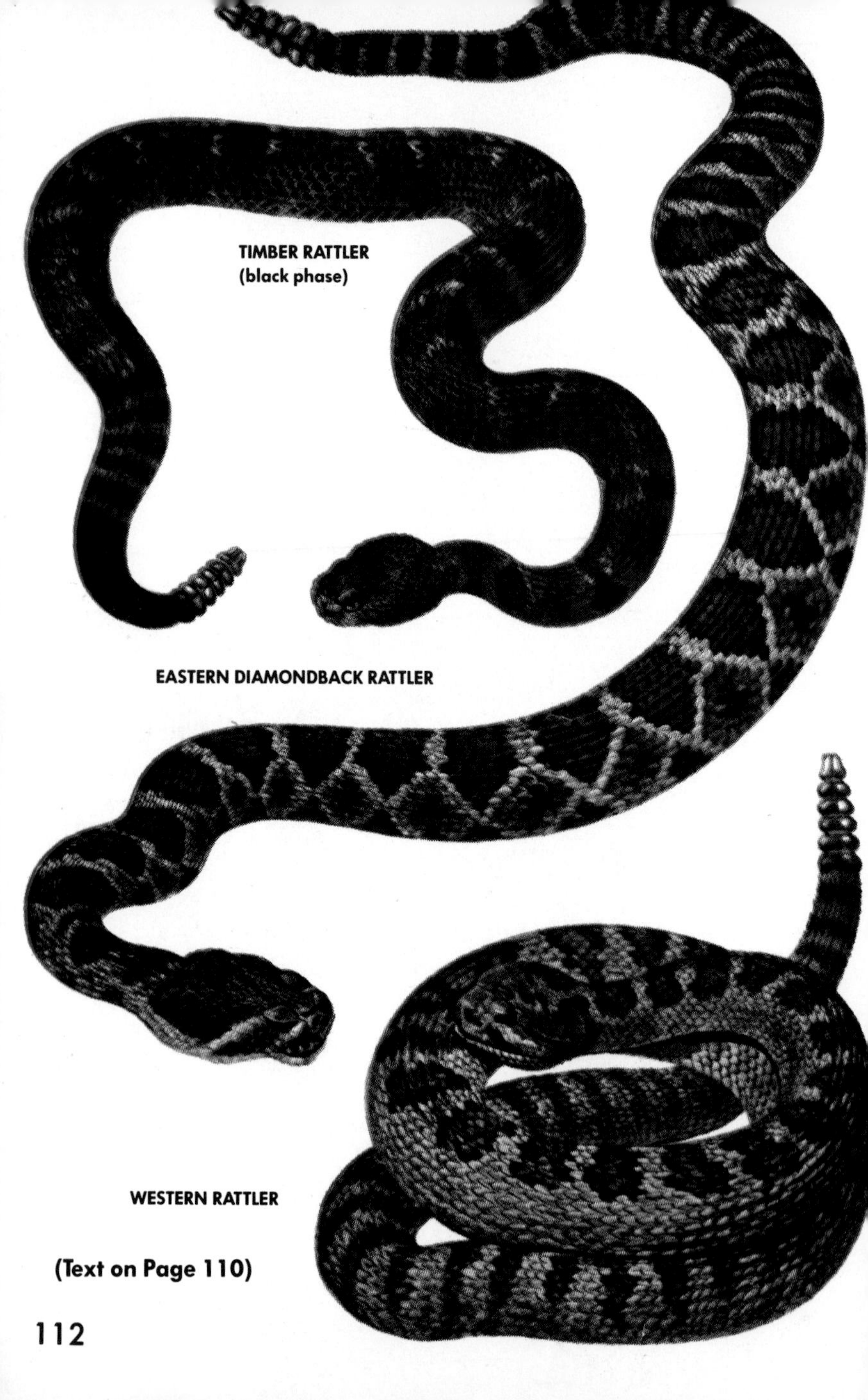

(Text on Page 110)

(Text on Page 110)

ALLIGATORS and CROCODILES form a distinct group of reptiles of ancient lineage. Once common in southern swamps, alligators have been reduced in number and range by hunters. Protection in recent years has resulted in recovery in some areas. Large specimens, 10 ft. and over, are still rare. Alligators are not usually dangerous. Reports of "man-eaters" usually refer to crocodiles of Africa or southern Asia. American Crocodile is paler than the alligator; its snout is pointed, narrower. Some of the teeth protrude, bulldog-fashion, from the sides of

AMERICAN CROCODILE

its jaw. Alligators and crocodiles feed on fish, turtles, birds, crayfish, crabs, and other water life. Both lay eggs, hatched by heat of the sun and of decaying vegetation. Crocodiles prefer salt marshes and even swim out into the ocean. Alligators prefer fresh water. Both are protected by law. Tropical American caimans now live in southern Florida. They are small (to 6 ft.) and have a bony ridge between front of orbits.

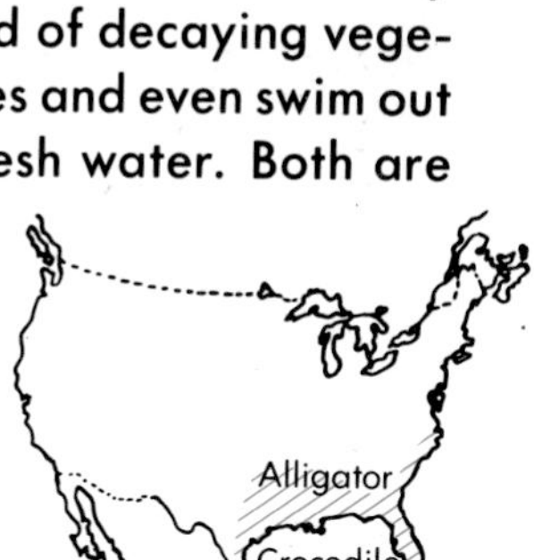

Alligator

Crocodile

AMPHIBIANS were the animals which, eons ago, first ventured out of water to live on land. Those that survive today are still poorly adapted to terrestrial life. Most spend at least part of their lives in water or in moist surroundings. Amphibians vary considerably in appearance, but all differ from reptiles in never having true clawed feet or a scaly epidermis. Of three groups, two are common. The salamanders and their kin are tailed amphibians. The frogs and toads are tailless when mature and often have hind legs better developed. The third group, the Caecilians—tropical, burrowing species—are living fossils, limbless, nearly tailless, earless, nearly or quite blind, earthwormlike.

Our amphibians lay jelly-covered eggs singly, in clumps, or in strings in quiet water or on moist leaf mold. In most species, these eggs hatch into larvae, or tadpoles, which usually breathe by means of gills and spend much or all of their life in water. Tadpoles feed mostly on microscopic plants and have mouth and digestive

Lizard Forefoot

Frog Forefoot

Salamander Forefoot

parts adapted for this diet. Larvae become air-breathing adults which may live in water or that live on land but return to the water to mate and lay eggs. Adults feed largely on insects. In the North they hibernate during winter underground or in mud on pond bottoms.

Frogs are divided into six major groups, as illustrated on page 6. The Tailed (p. 120) and Midwife Frogs represent the first. The second includes the Surinam and Mexican Burrowing Toad families, only the latter occurring in North America (extreme southern Texas). The Spadefoots (p. 121) and kin (two families) are the next group, followed by a large, diverse group of eleven families (three in North America; pp. 122-131) which includes toads, Treefrogs, and Chorus, Barking, and Chirping Frogs. Narrowmouth Toads and True Frogs (pp. 132-136), with two other non-American families, constitute the last two groups.

The three major groups of salamanders are redivided into nine families, of which eight occur in the United States. The group represented by the Slimy Salamanders and newts includes by far the largest number of species. The Tiger Salamanders and kin (pp. 144-145) are the only others that spend time on land.

American Toad Calling

FROGS and TOADS cannot be clearly distinguished, since there are many, not two, distinct groups in their order. True toads usually have rough or warty skins and live mainly on land. Frogs have smoother skins and live in water or wet places. Toads are plump, broad, and less streamlined than frogs. They are slower and cannot jump as well. Some frogs have such varied markings that identification is difficult. Added to this, the skin color and markings of some species change with their surroundings. Most male frogs and toads can inflate a sac in their throat when they make their characteristic sounds. There are about 82 species of tailless amphibians in this country. These fit into nine families, the largest of which are the treefrogs (pp. 124-129), the true toads (pp. 122-123), and the "true" frogs (pp. 132-135).

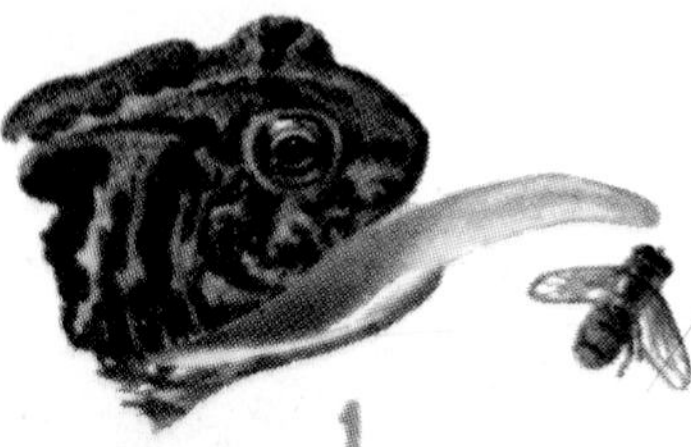

1

Extending Its Tongue

2

Catching Fly

Skin Color Changes with Surroundings

TADPOLES are the immature or larval stage of frogs and toads. The Barking Frogs (p. 130) are the only native frogs which do not have free-swimming tadpoles. Tadpoles are difficult to identify. The pictures may help you name some species. (See p. 133 for the Bullfrog tadpole.) Collect frogs' eggs or small tadpoles along the shores of ponds and ditches in spring; place them in an aquarium containing pond water and water plants. Do not overstock. As tadpoles hatch and begin to grow they will feed off bits of lettuce, which partly rots in the water. As your tadpoles change into frogs, provide a wooden float on which they can climb and rest.

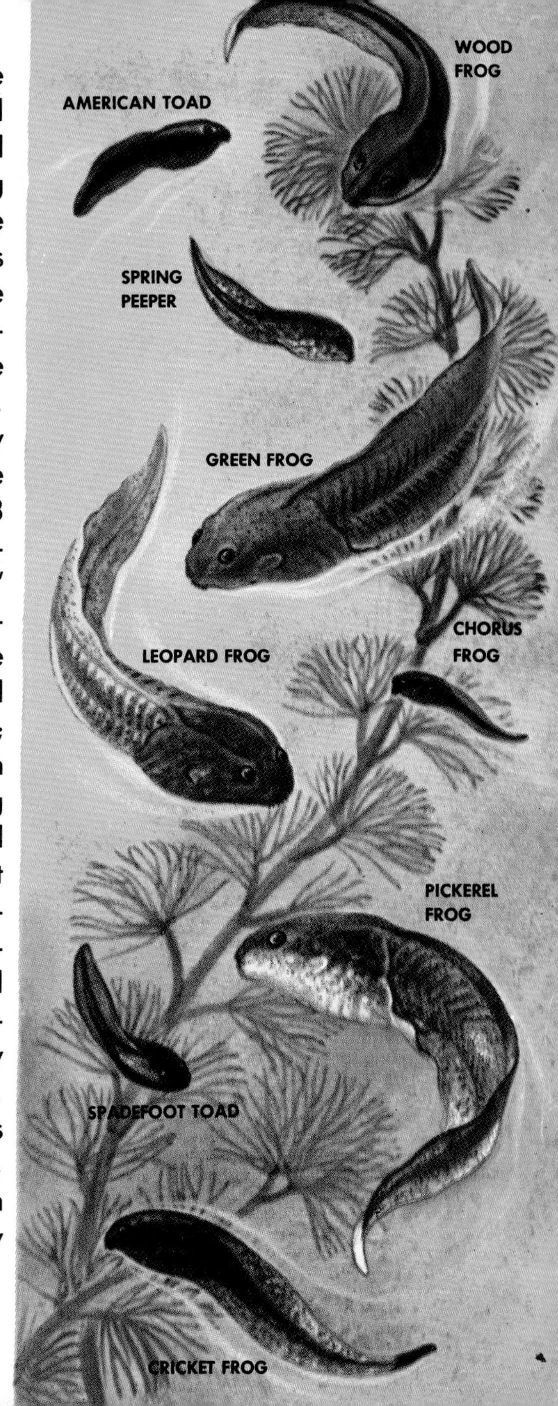

TAILED FROG is primitive. The male has a distinct tail-like organ. After breeding in late spring or early summer, strings of large eggs are found attached under rocks in rushing, cold mountain streams. Tadpoles cling to the rocks by means of a large sucking disc around the mouth. These small toads, 1 to 2 in. long, vary greatly in color, from gray and black to pink and brown. Note the webbed feet and the short, wide head with a light line sometimes across it.

WESTERN SPADEFOOT

EASTERN SPADEFOOT

SPADEFOOT TOADS (six species) have fleshy, webbed feet with large, horny, spadelike warts. In burrowing, the toad corkscrews backward and downward into the soil. It is found under logs or rocks, in shallow holes, coming out at night or after heavy rains to feed. Medium-sized (1½ to 3 in. long), it has relatively smooth skin. Eyes are large, with vertical pupils. Breeding is in late spring and early summer. Eggs are attached to plants at the water's edge.

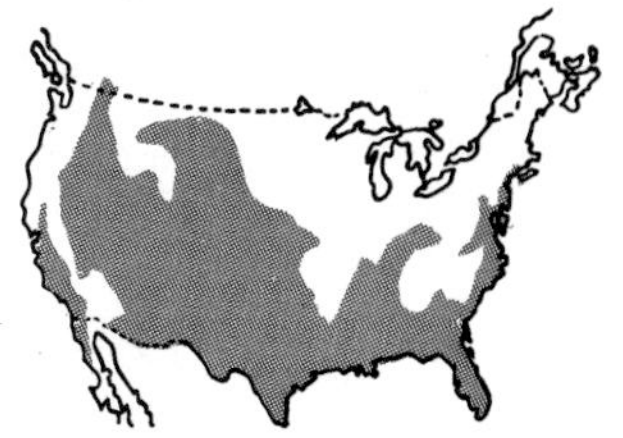

TRUE TOADS (18 species in U.S.), a much-maligned group of amphibians, were once wrongly credited with causing warts. Though clumsy, they are well adapted to life on land, feeding on almost any small, moving creature. They protect themselves by burrowing, playing dead, inflating their bodies, and exuding through their skin a white fluid which, in contact with eyes or mouth, is very poisonous. In breeding season and especially when it is raining, males make a species-characteristic trilling call.

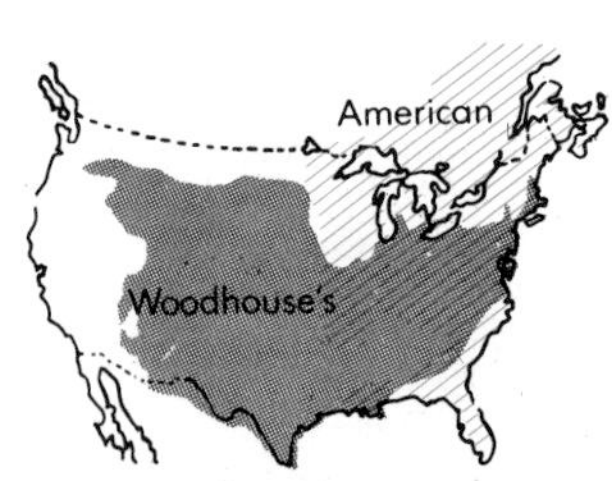

The American Toad is the common eastern species, 2 to 4 in. long. The male has a darker throat than the female. Fowler's Toad (the eastern race of Woodhouse's Toad) is more greenish and smaller, usually with smaller, more numerous warts and with a white line down the back. The Western Toad (2 to 5 in.) is very warty; the belly is mottled and the head more pointed than in eastern toads. The Great Plains Toad, commonly found along irrigation ditches and streams, is gray or brownish and somewhat varied in pattern.

SWAMP CHORUS FROG

CHORUS FROGS are seven species of small amphibians usually less than 2 in. long. They belong to the Treefrog family (though they rarely climb). All Chorus Frogs have slender bodies and pointed snouts. They breed early in spring, attaching small masses of eggs to leaves and stems in water. They are common at this time but later seem to disappear entirely, so their habits are not well known. They seldom climb more than a few inches above the ground; some cannot climb at all. The Striped Chorus Frog is small (¾ to 1½ in.), brownish or olive-colored, with distinct dark stripes on the back. The closely related Swamp Chorus Frog is common in

STRIPED CHORUS FROG

ORNATE CHORUS FROG

southern ditches and swamps. It is slender, olive green, with an even, granular skin. Spots on the back are irregular. The Ornate Chorus Frog completely lacks toe pads. It is chestnut brown with a dark mask and with dark spots on the sides; length 1 to 1¼ in. Strecker's Chorus Frog, a close relative of the Ornate, found farther west, is a stockier species (1 to ¾ in.), usually gray or greenish with darker spots and blotches on back and limbs. All either lack webs between toes and fingers, or have very short ones.

STRECKER'S CHORUS FROG

GRAY TREEFROG

GREEN TREEFROG

CANYON TREEFROG

PINE WOODS TREEFROG

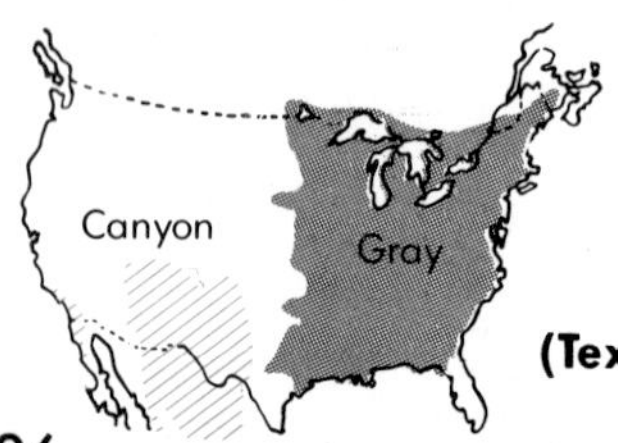

(Text on Page 128)

SQUIRREL TREEFROG

PACIFIC TREEFROG

SPRING PEEPER

BIRD-VOICED TREEFROG

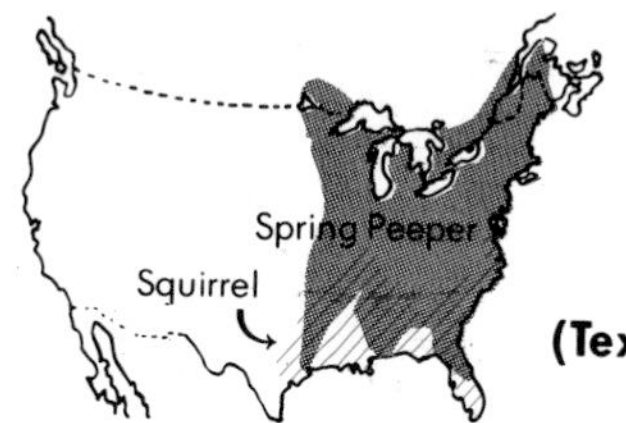

(Text on Page 128)

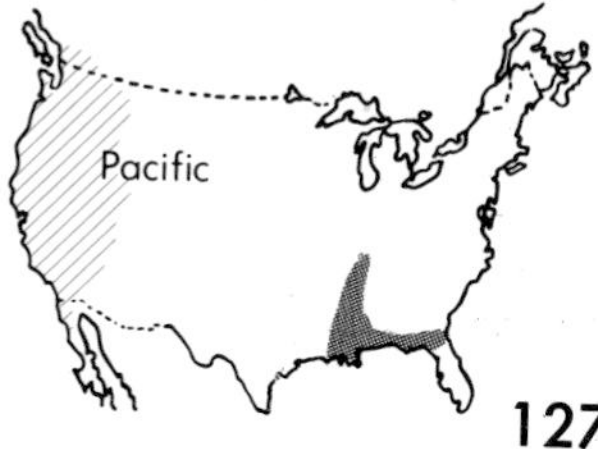

SPRING PEEPER

TREEFROGS comprise a large family, with seven genera in our area. True treefrogs, genus Hyla, are lightly built and live in trees and shrubs, clinging with the sticky pads on their toes. The skin, often slightly warty or rough, is usually brown or greenish. The call, heard in early spring, is loud, clear, musical. The frogs vary greatly in color and pattern, and can to a degree change color with their surroundings.

Gray Treefrog, with orange or brown thighs, back spotted or mottled gray or brown, skin slightly rough, is heard in midsummer in woods near water. Green Treefrog, most attractive, 1½ to 2½ in., with smooth green skin, slender and long-legged, has a penetrating honking call. Canyon Treefrog can change its color from brown or black to pale pinkish gray. The skin is rough. Eggs are laid singly, in water. Pine Woods Treefrog, legs brownish with small orange spots, irregular cross on back, ranges from greenish gray to reddish brown; found only in pine woods. Squirrel Treefrog, green to brown, usually spotted, skin smooth, has light stripe from eye to forelegs. Pacific Treefrog is gray, brown, or green; attractive; back sometimes spotted; brown V between eyes; 1 to 2 in.

Spring Peeper, our best-known eastern Treefrog, ¾ to 1¼ in., common in woodland swamps, is light brown or gray with dark diagonal cross on back. Bird-voiced Treefrog, dusky-colored, with greenish thigh and three rows of spots or a cross on the back, utters a unique whistle.

CRICKET FROGS are really small (¾ to 1½ in.) Treefrogs without toe pads. Hence they cannot climb. Color varies from brown to gray and green, with darker markings that may be brown or even reddish. A dark triangle is usually present atop the head. The skin is slightly rough. Cricket Frogs, common throughout the East, get their name from their call—a sharp, rapid metallic clicking. Eggs are laid singly, attached to plants in ponds and pools. Two species.

TEXAS BARKING FROG

FREE-TOED FROGS are a mostly tropical American family of many species. The Barking Frog lives in limestone ledges or caves. Eggs are laid in moisture-filled crevices. The tadpoles do not hatch but remain within the egg till they have developed into tiny frogs. Barking Frogs are short, squat, with wide, flat heads. The tiny Greenhouse species, imported from the West Indies, is only $\frac{3}{5}$ to $1\frac{1}{5}$ in. long. The larger Texas species (2 to $3\frac{1}{2}$ in.) has a call like a barking dog. All species are usually dark in color.

GREEN-HOUSE FROG

Tadpole in Egg (magnified 3 times)

WHITE-LIPPED and CHIRPING FROGS are really Mexican species. The first is a medium-sized, smooth-skinned frog (1½ to 2 in.), marked as its name indicates. It lays eggs in a frothy mass at the edge of ponds. Chirping Frog (three species) is smaller, with more pointed nose and granular skin. Its eggs, laid on land, hatch into legged frogs. There are no free-swimming tadpoles. This dull gray-green frog makes a faint whistling chirp.

Chirping
White-lipped

CRAWFISH FROG

CRAWFISH and RED-LEGGED FROGS introduce the "true" frog group—24 common species that have smooth, narrow bodies and long hind legs. The Crawfish Frog (2½ to 4½ in.), gray with small black spots, lives in the burrows of Gopher Tortoises or crayfish. Though fairly common, it is rarely seen. The large Red-legged Frog of the West (2 to 5 in.) is an even dark brown or olive above, colored below as its name indicates. It is a frog of moist forests, breeding in June or July. A new species has been discovered in Florida.

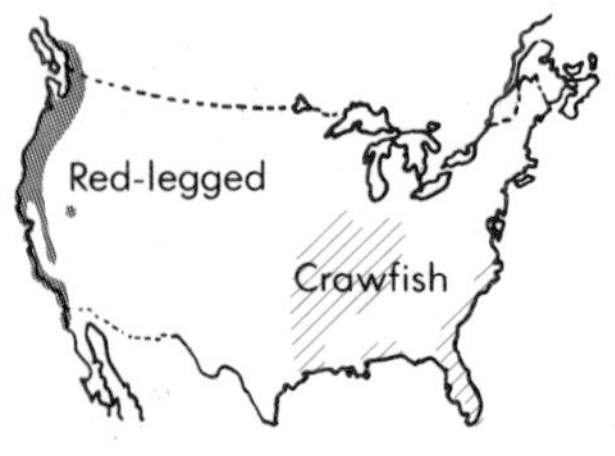

RED-LEGGED FROG

BULLFROG and GREEN FROG The Bullfrog is largest of our frogs (4 to 7½ in.). The male has very large "ears" (tympani) behind the eyes; the female's ears are smaller. The color is usually drab green. In the North, the large tadpole does not mature till the second year. The Green Frog is smaller (2 to 4 in.), with a yellowish throat, especially in the males. Both of these common frogs live in ponds and swamps. Both are solitary, laying eggs in spreading surface masses.

Bull
Green

BULLFROG Tadpole

PICKEREL FROG

NORTHERN LEOPARD FROG

PICKEREL and LEOPARD FROGS are common, attractive, and sometimes confusing. The former has square or rectangular spots on the back and reddish sides; legs are orange or reddish. Leopard or Meadow Frogs, a complex of seven to eight species all long thought to be one, have more rounded spots and greenish sides; legs are greenish. All are slender, smooth-skinned, and about 2 to 4 in. long. Found only about permanent water but may wander in moist, grassy meadows.

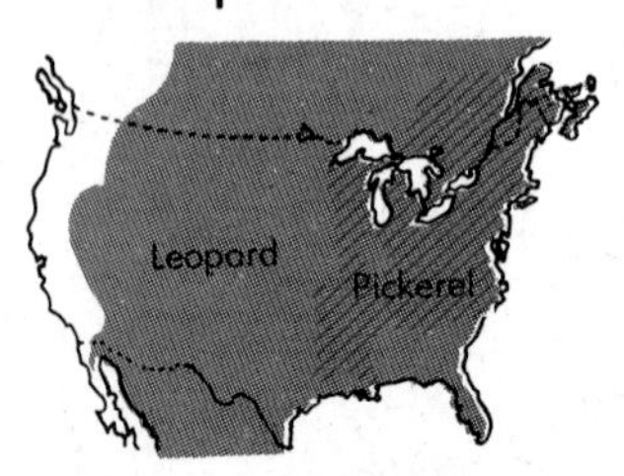

WOOD and SPOTTED FROGS The first is one of the most attractive common frogs: its fawn-brown skin is set off by a dark mask over the eyes. It prefers moist woods, breeds from May to July in woodland pools. Eggs are laid near shore in rounded mass, 2 to 4 in. across, containing 2,000 to 3,000 individual eggs. Length: 1½ to 3 in. The Spotted Frog (3 to 4 in.) is a western species typical of mountain areas. It is dark brown, sometimes spotted with skin slightly roughened. A light streak marks the edge of the upper jaw.

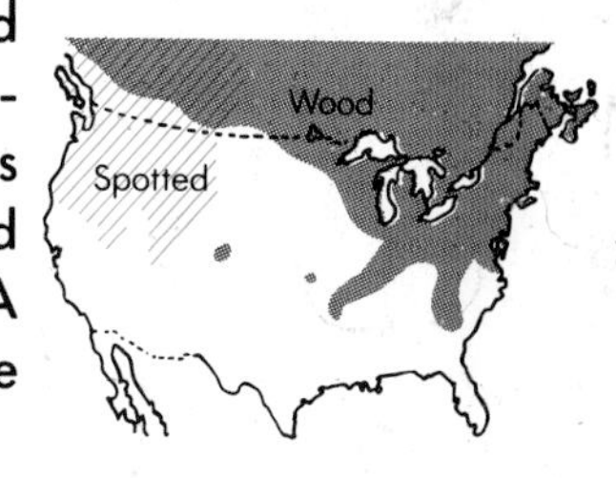

SPOTTED FROG

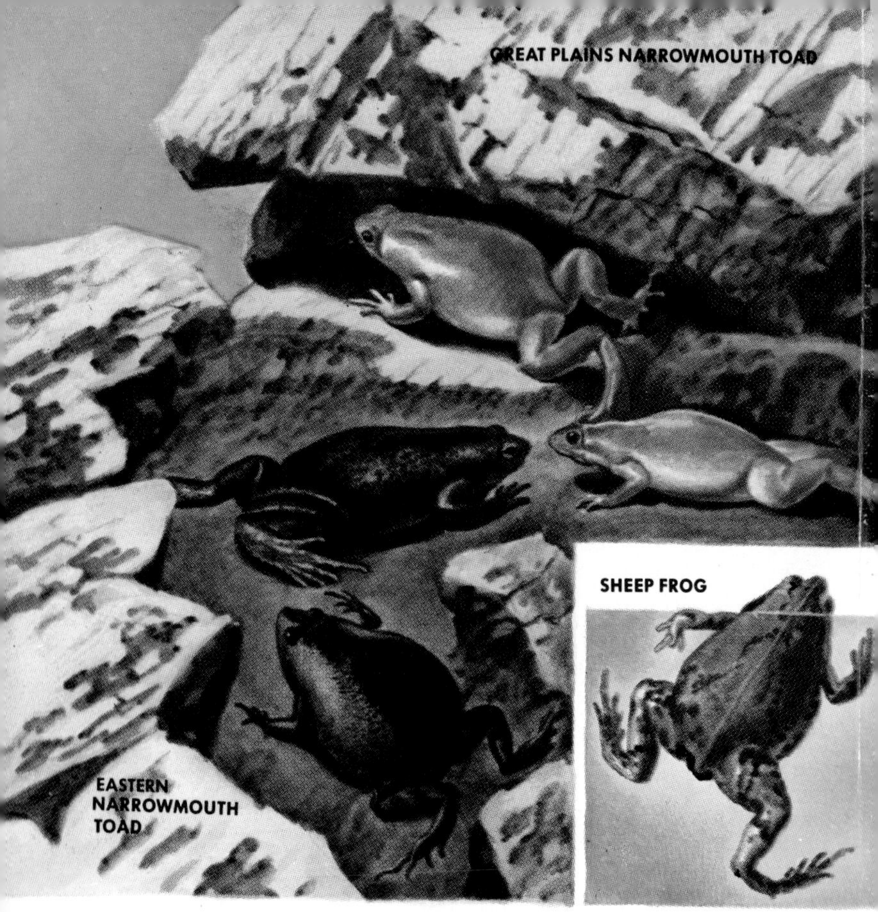

GREAT PLAINS NARROWMOUTH TOAD

SHEEP FROG

EASTERN NARROWMOUTH TOAD

NARROWMOUTH TOADS have small, wedge-shaped heads with a fold of skin crossing the head just back of the eyes. They are dark or mottled; undersides lighter. Nocturnal toads with tiny voices, they often hide under logs and rocks. The Sheep Frog, a related species, has narrow head but loose, dark skin, with a narrow yellow or orange stripe down the back. It breeds (March-September) in shallow ponds or large rain-water pools.

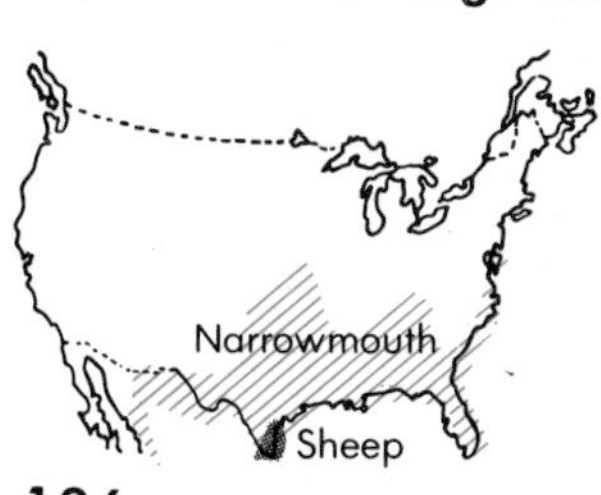

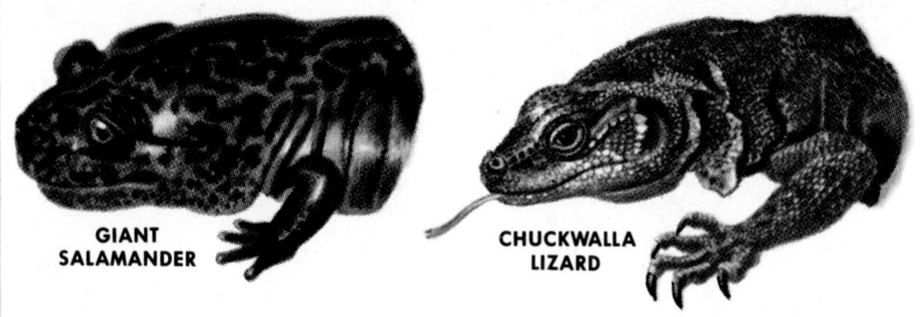

SALAMANDERS are tailed amphibians. About 118 species, in eight families, are found in this country. They differ from lizards (pp. 44-45) in lacking a scaly skin and claws. Salamanders never have more than four toes on the front feet; lizards usually have five. Nocturnal, all avoid direct sun. During the breeding season they move about more and hence are more likely to be seen. Some spend their entire lives in water; others live on moist land, some returning to water to mate and lay eggs. The eggs, with a jellylike coating, are laid singly, in strings or in small clumps. Some terrestrial salamanders have no free-living larval stage. Salamanders may be kept in terraria like frogs and toads or in aquaria like fish, depending on kind and life-stage. Feed them live insects.

Eggs of Hellbender

Eggs of Dwarf Salamander

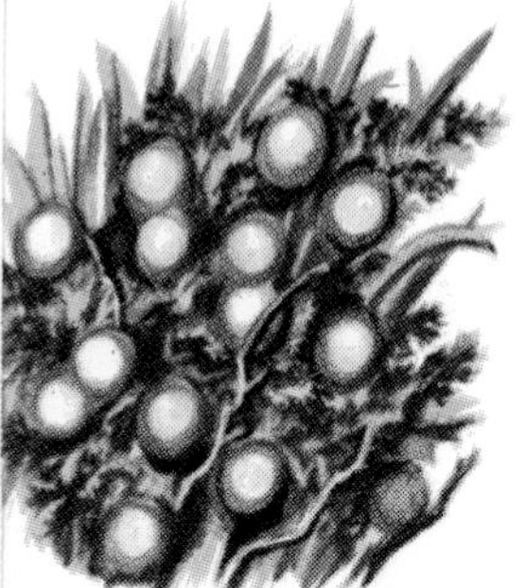

Eggs of Spotted Salamander

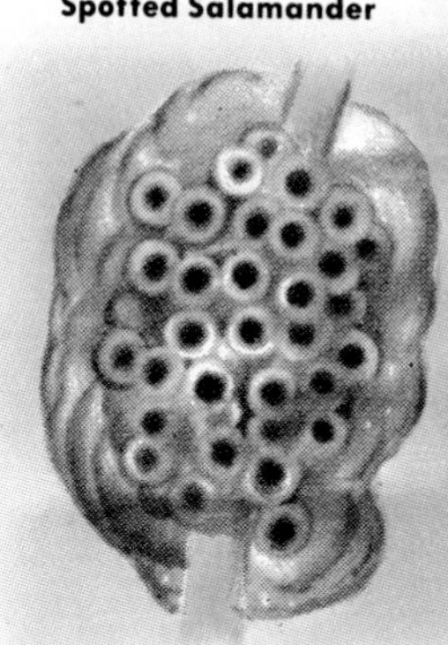

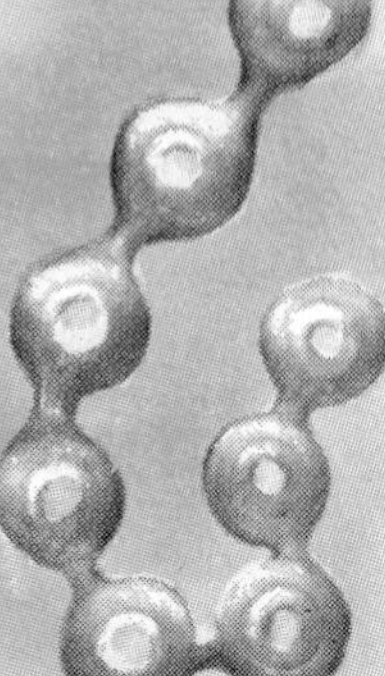

MUDPUPPY

MUDPUPPY and WATERDOGS are large (12 in.) aquatic salamanders of rivers and lakes. Color varies—often dark brown above, paler on belly with dark spots. Larvae throughout life, they have bushy red gills. Eggs are laid in late spring attached to rocks under water. The eggs hatch in 40 to 60 days. Hatchlings, striped on their back and sides, are about an inch long; they mature in about five years. Five species occur in the United States.

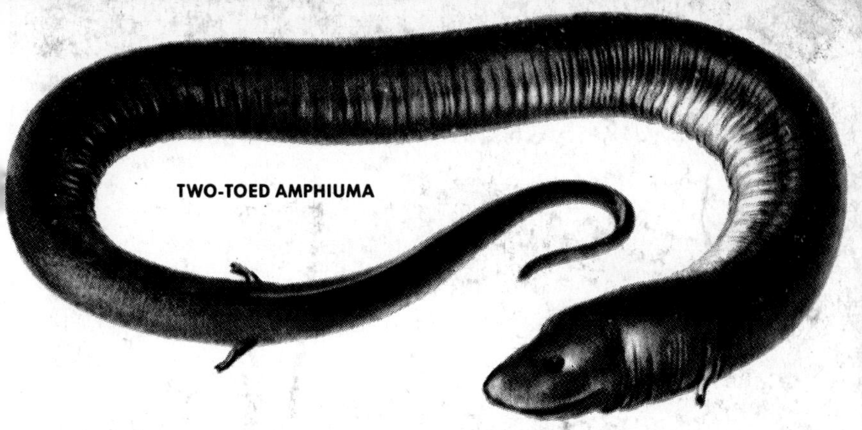

AMPHIUMAS and HELLBENDER are large aquatic salamanders. The former (three species), smooth and eel-like, grow 30 to 36 in. long, with four tiny, useless, one- to three-toed feet. They are often found in ditches, in burrows, or under debris. The female lays a mass of eggs under mud or rotted leaves. She may remain near to guard them. The Hellbender (16 to 20 in.) is shorter and broader, and lives farther north. Its wrinkled skin makes identification easy. The color varies from spotted yellowish to red and brown. Eggs are laid under rocks in shallow water.

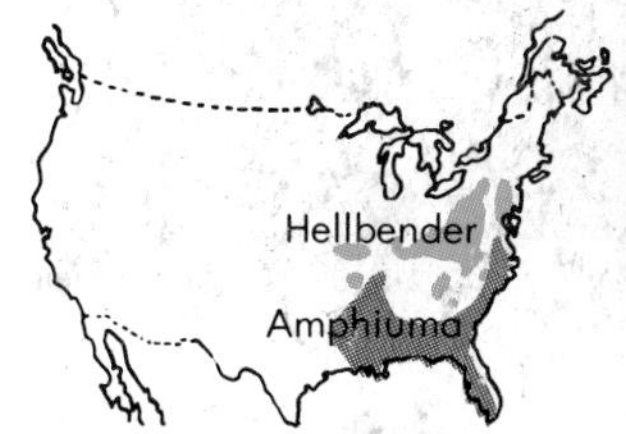

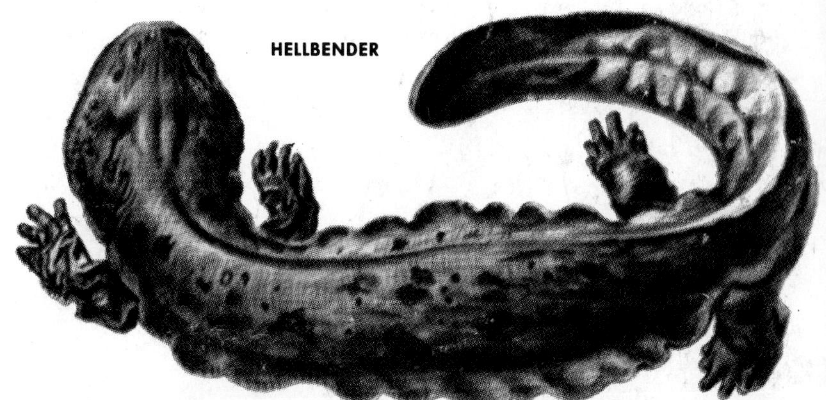

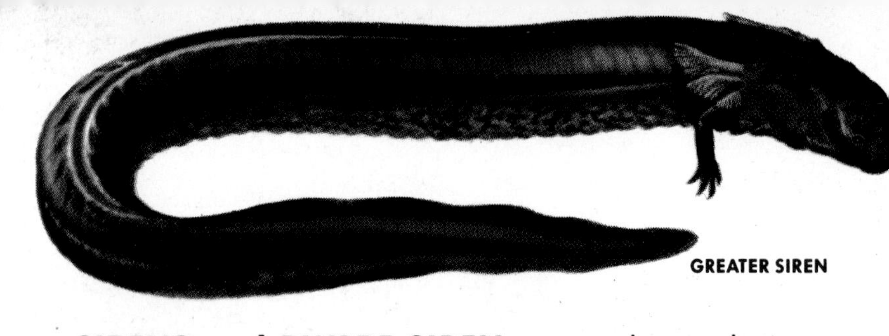

GREATER SIREN

SIRENS and DWARF SIREN are southern salamanders of rivers, swamps, and ponds. Both have external gills and both have only front legs. The Sirens (two species) are larger (about 30 in.), gray, olive green, or blackish with spots and blotches. The Dwarf Siren (one species), 5 to 8 in. long, has smaller gills and legs; it occurs in southern streams and waterways. Light stripes down the back and sides are a characteristic marking. Both feed on insects, worms, larvae, and other small water animals.

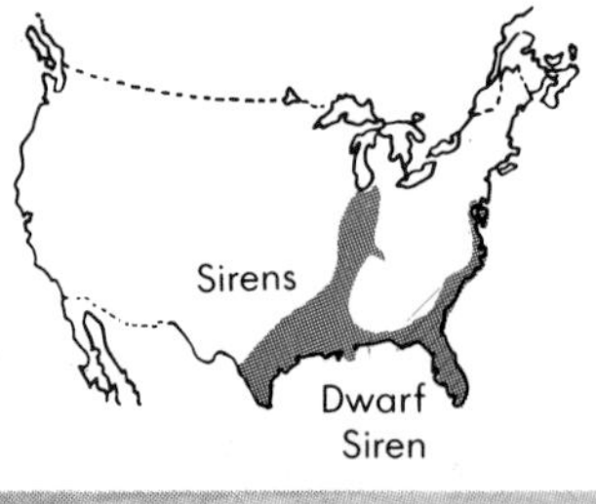

Sirens

Dwarf Siren

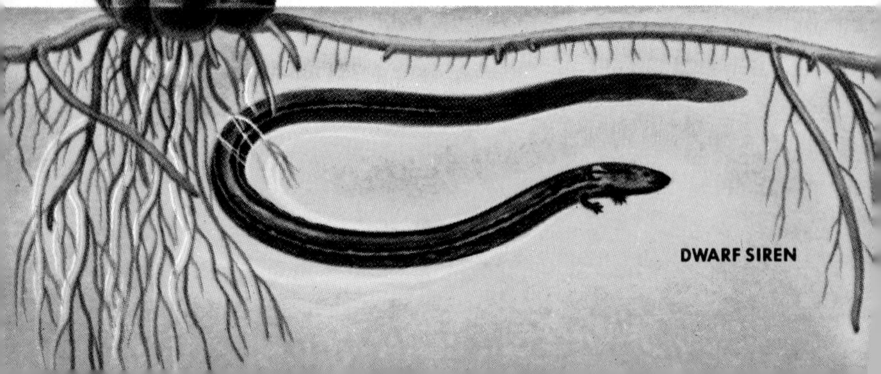

DWARF SIREN

GIANT and OLYMPIC SALAMANDERS are two northwestern genera, in separate families. The first (three species; 9 to 12 in.), includes our largest land species. They are found on moist slopes under rocks and logs. Larvae live in nearby streams. The back color varies—usually mottled; legs darker. Olympic Salamander, smaller (3½ in.), prefers the humid coastal coniferous forests, where it is usually found in or along clear streams.

Olympic
Giant
(red)

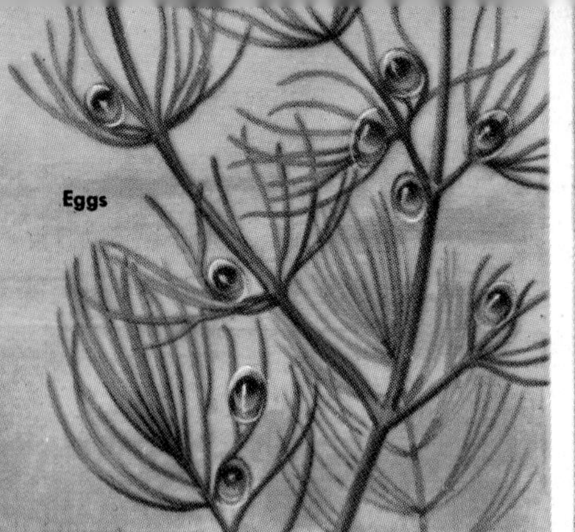

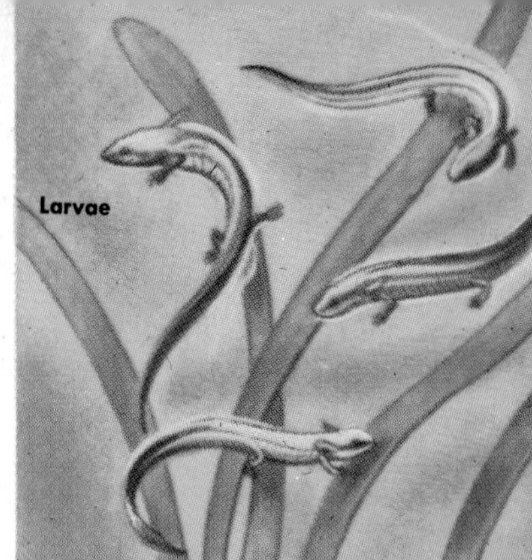

Stages in Life Cycle of the Eastern Newts

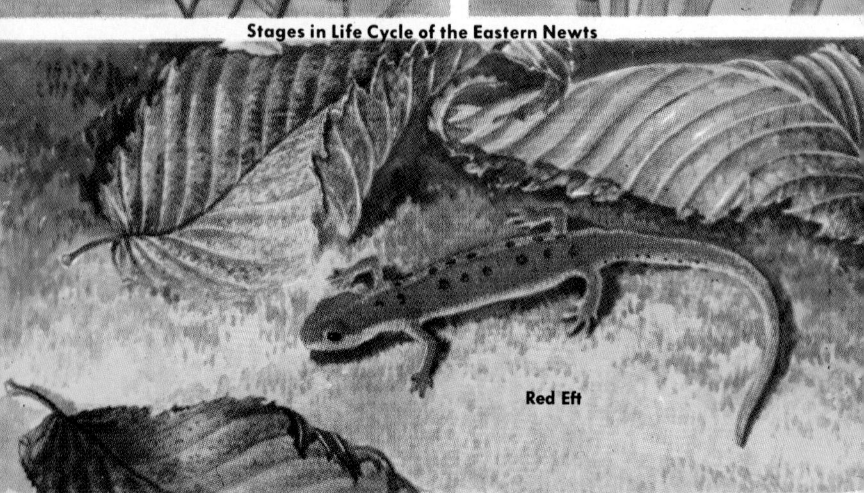

NEWTS are attractive, interesting salamanders. Of the six species, the eastern (three species, 3 in. long) are perhaps the best known. Their eggs, laid in spring, on stems and leaves of water plants, hatch into larvae. After three or four months in the water these usually leave to spend two or three years on land as an unusual form, known as the Red Eft. When the Efts return permanently to water, they change color and develop a broad swimming tail. Some newts skip the eft stage. Newts feed on worms, insect larvae, and small aquatic

animals, and are easy to care for in captivity. Red Efts, fed on live insects, do well in terraria. Adults thrive in aquaria, feeding on small bits of liver or other meat. The Pacific Newts (three species) are about twice the size of the eastern species and differ in appearance too. Adults are land-dwellers, returning to water only to breed. They are reddish or dark brown, belly much lighter yellow or orange. Found in moist woods and mountain ponds.

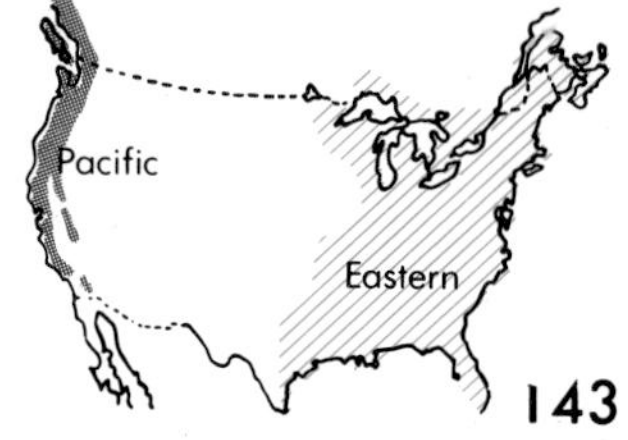

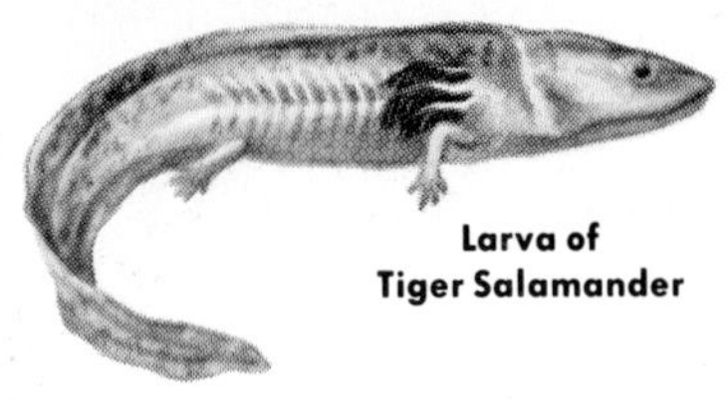

Larva of Tiger Salamander

SPOTTED SALAMANDER (7 in.) has large, round, yellow or orange spots on a black skin. Like others in this group (16 species) it has vertical grooves on its sides. It is found in moist woods; breeds in ponds and temporary pools. Adults migrate considerably, returning to water to breed. They feed on worms, grubs, and insects.

TIGER SALAMANDER (8 in.) is like the Spotted, but the spots, when present, are larger, more irregular, and extend down the sides and onto the belly. Some larvae do not develop into the land form; they spend their entire life in water, where they eventually breed. Tiger Salamanders are known to live over ten years.

MARBLED SALAMANDER (4 in.) is smaller than others in this group, but like most is a stout, thick-set creature. Variable markings on the black skin, white on males, grayish on females, in irregular fused bands. The larvae are a mottled brown.

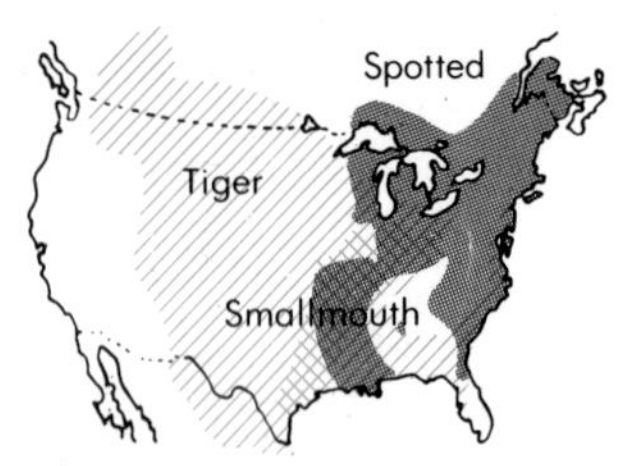

JEFFERSON SALAMANDER is slender (6½ in.); also called Blue-spotted, for the markings on its brownish skin. It lives in woods along swamps and streams. Trunk and tail have vertical grooves.

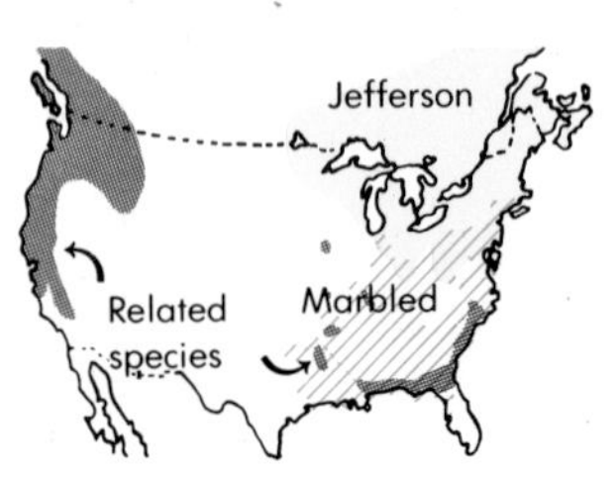

SMALLMOUTH SALAMANDER (5½ in.) is found in varying habitats from swampy lowlands to upland woods. It is a burrower beneath logs and rocks near streams. The color is a faintly blotched slate gray or brown, lighter beneath.

SPOTTED SALAMANDER
TIGER SALAMANDER
MARBLED SALAMANDER
JEFFERSON SALAMANDER
SMALLMOUTH SALAMANDER

DUSKY SALAMANDERS include 11 species of average-sized (3½ in.), inconspicuous salamanders with highly variable color and pattern. Their dark, mottled skins blend with rocks and moss along streams where they live. The sides are grooved vertically. Note a small light bar from eye to corner of mouth. The Mountain, Pigmy, and Seal species differ from the Dusky in having a light, irregularly spotted band down the back.

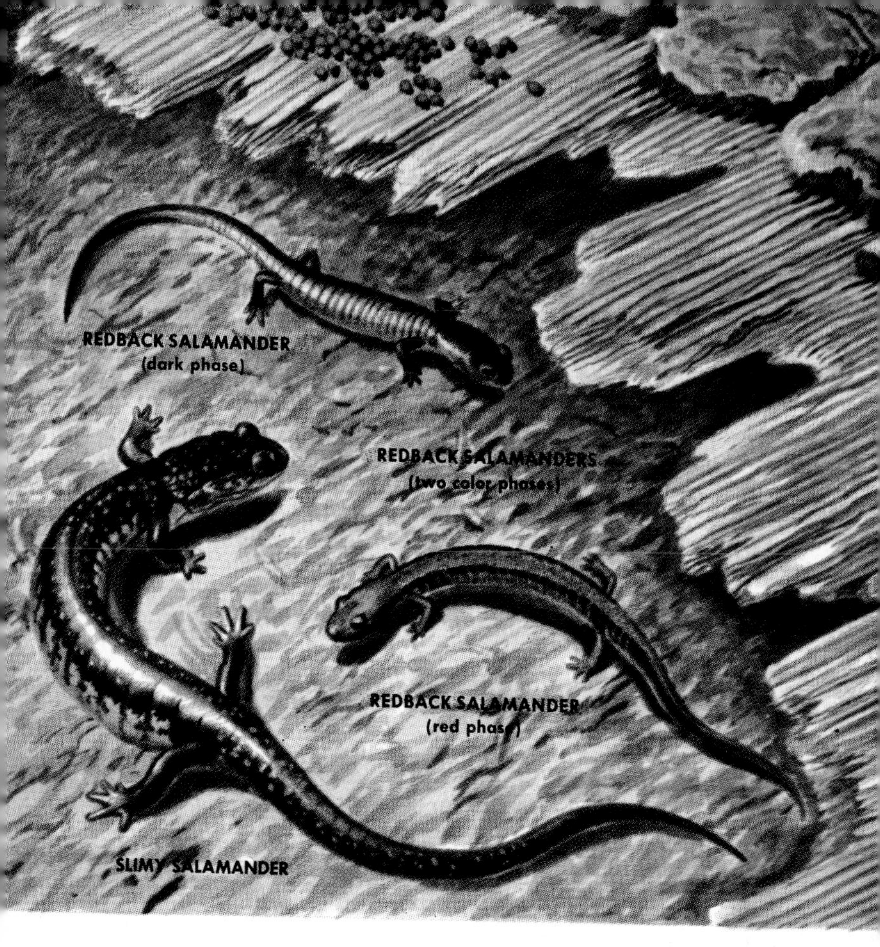

REDBACK and SLIMY SALAMANDERS are land species of our largest group (29 species). Often found in leaf mold and under rotted logs, both breed on land and lay eggs in moist nests in rotted bark or logs. Redback (3 in. long) has two color phases; only one has the red stripe down the back. Slimy Salamander, larger (6 in.), has blue-black skin with small, irregular light spots on the back, and a grayish belly.

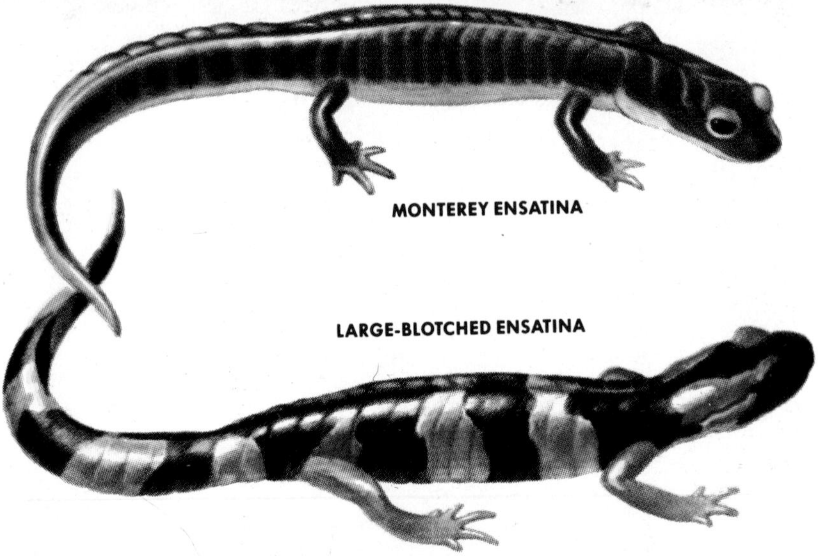

MONTEREY ENSATINA

LARGE-BLOTCHED ENSATINA

ENSATINA and SLENDER SALAMANDERS are western species. The single species of Ensatina (seven races) varies from black to red, usually with red or yellow-orange blotches. These medium-sized salamanders (4 to 5 in.) occur in the mountains, in oak and evergreen forests. They exhibit an unusual, complex courtship pattern. Slender Salamanders (ten species; 4 in.) are thin and wormlike, with a long tail. Color is dark, often spotted or streaked. They are found under rotted logs or leaves where they lay their eggs, from which tiny miniatures of the adults emerge.

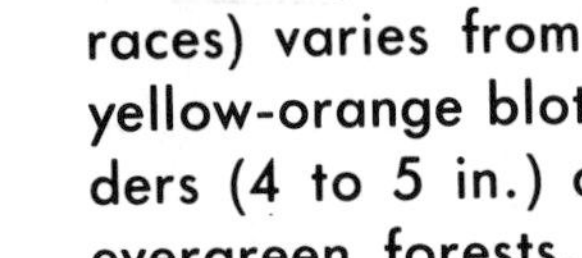

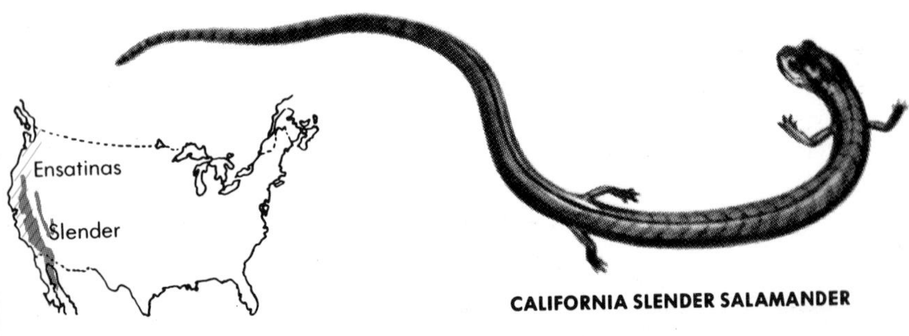

CALIFORNIA SLENDER SALAMANDER

ARBOREAL SALAMANDER

CLIMBING SALAMANDERS (five species; 4 in. long) live on opposite sides of the country. Green Salamander is found on the rocky hillsides of the Appalachians, under logs or in crevices of rocks. It is dark, with greenish blotches. The Arboreal Salamander of the Pacific Coast frequently lives in water-soaked cavities of trees. Sometimes a whole colony is found in one of these holes, where eggs are laid, also. Arboreal Salamanders also live on the ground, under logs, rocks, and bark. Their color is light brown, paler below with few if any markings. Males have long teeth at front of upper jaw.

GREEN SALAMANDER

GROTTO SALAMANDER

TEXAS BLIND SALAMANDER

BLIND SALAMANDERS are unusual animals found only in deep wells and underground streams of caves. They are a pale yellowish in color, with eyes reduced in size or completely undeveloped. The larvae of the Grotto Salamander (adults 3¾ in.), found in open streams, have dark-colored skins and normal eyes. The young of the two Texas species (adults 4 in.) resemble the pale adults. Another rare Blind Salamander has been found in Georgia and Florida.

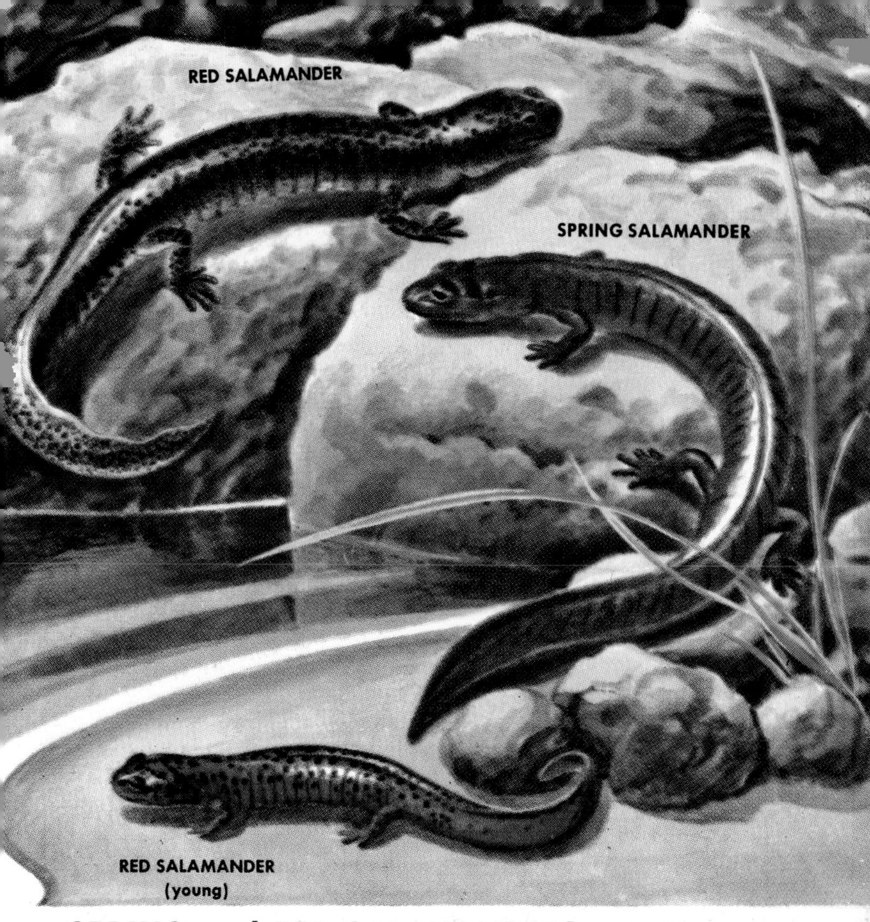

SPRING and RED SALAMANDERS sometimes prey upon other salamanders. The Spring Salamanders (three species; 5 in. long) are brownish or reddish brown, with vague spots or blotches. Young adults, newly transformed from larvae, are brighter red. This is also true of Red Salamanders (two species; 5 in.). Young adults are bright red with small dark spots; older ones, dull and darker. Both of these salamanders are commonest in hilly or mountain areas along streams or near ponds.

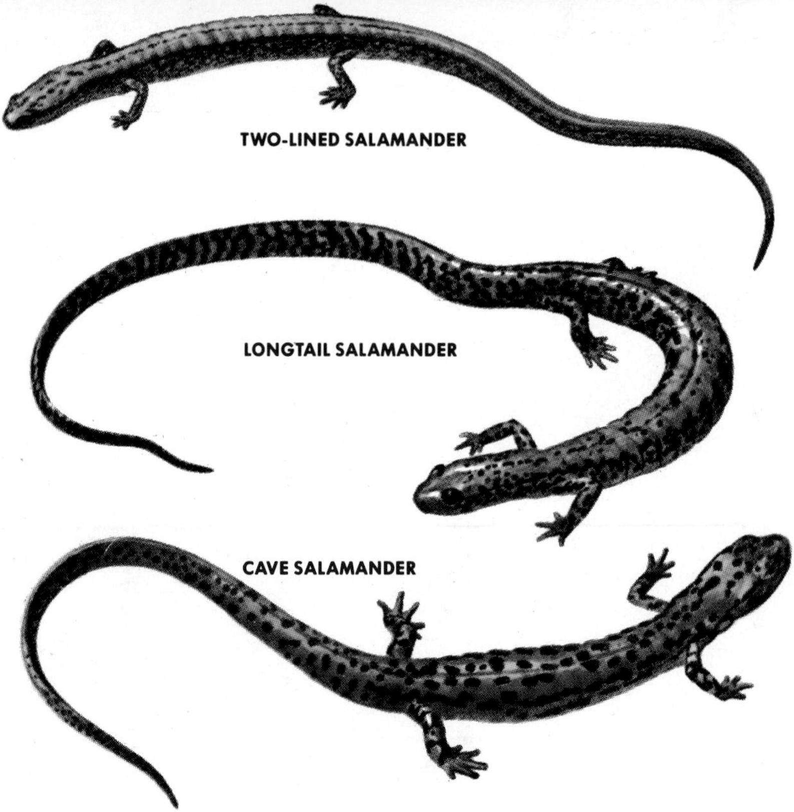

TWO-LINED SALAMANDER

LONGTAIL SALAMANDER

CAVE SALAMANDER

TWO-LINED, LONGTAIL, and CAVE SALAMANDERS represent a common but inconspicuous group (ten species). Two-lined (3 in.) is so marked, with a broken row of dark dots between the lines on its sides. Longtail (about 5 in.) is thin, yellow to orange, with dark tail bars. Both prefer moist sites under logs and rocks, though the Two-lined also prefers brooksides. The Cave Salamander (5 in.) is seen near the entrances of caves and under moist, overhanging rocks. Color is variable, usually yellow or orange with scattered black spots.

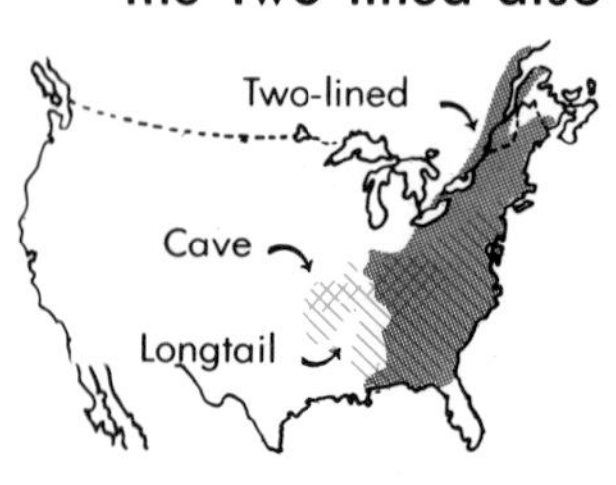

FOUR-TOED SALAMANDER is so called because both front and hind feet are four-toed. It is one of the smallest salamanders (2½ in.), fairly common in wooded areas, swamps, and bogs. The dull red-brown back is mottled with darker patches; the belly is lighter, with brown spots. Males are smaller than females and have longer tails. The female lays her eggs in a mossy cavity and stays with them till they hatch, in about two months. The larvae leave the water in six weeks to complete their development on land. They mature in about two years.

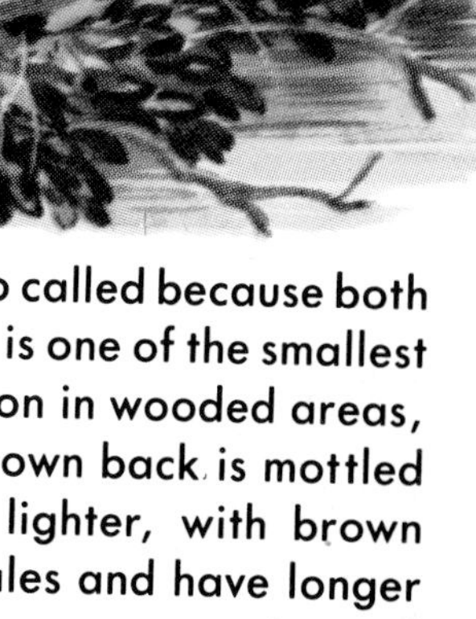

BOOKS FOR FURTHER STUDY

Catalogue of American Amphibians and Reptiles. Society for the Study of Amphibians and Reptiles, c/o Publications Secretary, Dr. Douglas H. Taylor, Dept. Zoology, Miami Univ., Oxford, Ohio, 45056. The most definitive digest of all known information on American (mostly North American) herps. About half complete in 1986.

Conant, Roger. *A Field Guide to Reptiles and Amphibians of Eastern and Central North America*. Houghton Mifflin, Boston, 1975. Complete field guide coverage of herps east of the Rocky Mountains.

Duellman, William E. and Linda Trueb. *Biology of Amphibians*. McGraw-Hill, New York, 1986. A concise, scholarly summary of all current knowledge about amphibians of the world.

Ernst, Carl H., and Roger W. Barbour. *Turtles of the United States*. Kentucky University Press, Lexington, 1972. The best recent account of the natural history of North American turtles, very thorough. No comparable account exists for other herps.

Goin, Coleman J., Olive B. Goin, and George R. Zug. *Introduction to Herpetology*. Third edition. Freeman, San Francisco, 1978. Herpetological fundamentals, North American emphasis.

Oliver, James A. *The Natural History of North American Amphibians and Reptiles*. Van Nostrand Reinhold, New York, 1955. Although old, still the best scholarly summary of the subject.

Smith, Hobart M., and Edmund Brodie, Jr. *Reptiles of North America, A Golden Field Guide*. Golden Press, New York, 1982. How to recognize—plus fascinating facts about—all the reptiles of the U.S. and Canada, with simplified identification keys and range maps.

Stebbins, Robert C. *A Field Guide to Western Reptiles and Amphibians*. Houghton Mifflin, Boston, 1985. The companion volume for Conant's guide, covering the area west from the Rocky Mountains.

ZOOS, MUSEUMS, AND STUDY COLLECTIONS

Washington, D.C.: U.S. National Museum, National Zoological Park. **New York City**: American Museum of Natural History, Staten Island Zoo, Bronx Park Zoo. **Chicago, Ill.**: Field Mus. of Nat. Hist., Brookfield Zoo, Lincoln Park Zoo. **Cambridge, Mass.**: Harvard Museum of Comparative Zoology. **Philadelphia, Pa.**: Philadelphia Zoological Park. **Los Angeles, Calif.**: Los Angeles County Museum. **Ann Arbor, Mich.**: Univ. of Mich. Museum of Zoology. **Lawrence, Kans.**: Univ. of Kansas Museum of Natural History. **San Antonio, Texas**: San Antonio Zoo. **San Diego, Calif.**: Zoological Park. **Berkeley, Calif.**: Univ. of Calif. Mus. of Vert. Zoology. **St. Augustine, Fla.**: St. Augustine Alligator Farm. **Rapid City, S. Dak.**: Black Hills Reptile Gardens.

SCIENTIFIC NAMES

The scientific names of illustrated reptiles and amphibians follow. Heavy type indicates pages where they appear. The genus name is first, then the species. A third name is the subspecies. If the genus name is abbreviated, it is the same as the genus name given just before it.

20 Leatherback: Dermochelys coriacea
Hawksbill: Eretmochelys imbricata
21 Loggerhead: Caretta caretta
Green: Chelonia mydas
22 Sternotherus odoratus
23 Eastern: Kinosternon subrubrum subrubrum
Yellow: K. flavescens
24 Chelydra serpentina
25 Macroclemys temminckii
26 Trionyx spiniferus
27 Gopherus polyphemus
28–29 Trachemys scripta elegans
30 Pseudemys concinna
31 Deirochelys reticularia
32 Eastern: Chrysemys picta picta
Southern: C. picta dorsalis
Western: C. picta bellii
33 Chrysemys picta marginata
34 Graptemys kohnii
35 Graptemys geographica
36 Emydoidea blandingii
37 Malaclemys terrapin
38–39 Eastern: Terrapene carolina
Western: T. ornata
40 Clemmys guttata
41 Clemmys marmorata
42 Clemmys muhlenbergi
43 Clemmys insculpta
46 Leaf-toed: Phyllodactylus xanti
Dwarf: Sphaerodactylus cinereus
Turkish: Hemidactylus turcicus
47 Coleonyx variegatus
48 Anole: Anolis carolinensis
Chameleon: Chameleo vulgaris
49 Sauromalus obesus
50 Dipsosaurus dorsalis
51 True: Iguana iguana (juv.)
Spiny: Ctenosaura pectinata (juv.)
52 Crotaphytus collaris
53 Gambelia wislizenii
54 Tree: Urosaurus ornatus
Side-blotched: Uta stansburiana
55 Lesser Earless: Holbrookia maculata
Zebratail: Callisaurus draconoides
Fringe-toed: Uma notata
56 Texas Spiny: Sceloporus olivaceus
Western Fence: S. occidentalis
Sagebrush: S. graciosus
Crevice Spiny: S. poinsettii
Desert Spiny: S. magister
57 Sceloporus undulatus
58 Desert: Phrynosoma platyrhinos
Short-horned: P. douglassii
59 Phrynosoma cornutum
60 Granite: Xantusia henshawi
Arizona: X. vigilis arizonae
61 Eumeces obsoletus
62 Broadhead: Eumeces laticeps
Western: E. skiltonianus
Gilbert's: E. gilberti
Great Plains: E. obsoletus
63 Eumeces fasciatus
64 Ground: Scincella lateralis
Sand: Neoseps reynoldsi
65 Six-lined Racerunner: Cnemidophorus sexlineatus
Western Whiptail: C. tigris
66 Texas: Gerrhonotus liocephalus
Western: Elgaria multicarinata
67 Ophisaurus ventralis
68 Worm: Rhineura floridana
Legless: Anniella pulchra
69 Heloderma suspectum
72 Leptotyphlops humilis
73 Rubber: Charina bottae
Rosy: Lichanura trivirgata

74 Farancia erytrogramma
75 Farancia abacura
76 Northern: Diadophis punctatus edwardsii
77 Rough: Opheodrys aestivus
Smooth: O. vernalis
78 Rough Earth: Virginia striatula
Ground: Sonora semiannulata
Short-tailed: Stilosoma extenuatum
Western Shovelnose: Chionactis occipitalis
79 Black Swamp: Seminatrix pygaea
Striped Crayfish: Liodytes alleni
Sharptail: Contia tenuis
Banded Sand: Chilomeniscus cinctus
80 Chilomeniscus cinctus
81 Western: Heterodon nasicus
Eastern: H. platyrhinos
82 Vine: Oxybelis aeneus
Western Hooknose: Gyalopion canum
Cateyed: Leptodeira septentrionalis
83 Southeastern Crowned: Tantilla coronata
Pine Woods: Rhadinaea flavilata
Mexican Hooknose: Ficimia streckeri
Worm: Carphophis amoena
Black-striped: Coniophanes imperialis
85 Western Yellowbelly: Coluber constrictor mormon
Northern Black: C. constrictor constrictor
86 Masticophis flagellum flagellum
87 Masticophis taeniatus taeniatus
88 Salvadora grahamiae lineata
89 Phyllorhynchus browni
90 Gray: Elaphe obsoleta spiloides
Yellow: E. obsoleta quadrivittata
91 Elaphe obsoleta obsoleta
92 Elaphe guttata guttata
93 Elaphe vulpina
94 Drymarchon corais
95 Arizona elegans
96 Pituophis melanoleucus melanoleucus
97 Pituophis melanoleucus sayi
98 Scarlet: Lampropeltis triangulum elapsoides
Eastern Milk: L. triangulum triangulum
Eastern Kingsnake: L. getulus getulus
99 Speckled: Lampropeltis getulus holbrooki
Calif.: L. getulus californiae
100 Cemophora coccinea
101 Rhinocheilus lecontei
102 Northern: Nerodia sipedon sipedon
Redbelly: N. erythrogaster erythrogaster
103 Green: Nerodia cyclopion
Diamondback: N. rhombifera
104 Plains: Thamnophis radix
Western Terrestrial: T. elegans
105 Eastern Ribbon: Thamnophis sauritus
Common: T. sirtalis
106 Lined: Tropidoclonion lineatum
Brown: Storeria dekayi
Redbelly: S. occipitomaculata
107 Night: Hypsiglena torquata
Lyre: Trimorphodon biscutatus lambda
108 Eastern: Micrurus fulvius
Western: Micruroides euryxanthus
109 Copperhead: Agkistrodon contortrix
Cottonmouth: A. piscivorus
111 Pigmy: Sistrurus miliarius
Massasauga: S. catenatus
112 Timber: Crotalus horridus
Eastern Diamondback: C. adamanteus
Western: C. viridis
113 Sidewinder: Crotalus cerastes
Western Diamondback: C. atrox
Red Diamond: C. ruber

114 Alligator mississippiensis
115 Crocodylus acutus
118 American Toad: Bufo americanus
120 Ascaphus truei
121 Western: Scaphiopus hammondii
Eastern: S. holbrookii
122 American: Bufo americanus
Fowler's: B. woodhousii fowleri
123 Western: Bufo boreas
Great Plains: B. cognatus
124 Southern: Pseudacris nigrita
Striped: P. triseriata
125 Ornate: Pseudacris ornata
Strecker's: P. streckeri
126 Gray: Hyla versicolor
Green: H. cinerea
Canyon: H. arenicolor
Pine Woods: H. femoralis
127 Squirrel: Hyla squirella
Pacific: H. regilla
Spring Peeper: H. crucifer
Bird-voiced: H. avivoca
129 Acris crepitans
130 Texas: Hylactophryne latrans
Greenhouse: Eleutherodactylus planirostris
131 White-lipped: Leptodactylus fragilis
Cliff Chirping: Syrrhophus marnockii
132 Crawfish: Rana areolata
Red-legged: R. aurora
133 Green: Rana clamitans
Bullfrog: R. catesbeiana
134 Pickerel: Rana palustris
Northern Leopard: R. pipiens
135 Wood: Rana sylvatica
Spotted: R. pretiosa
136 Great Plains: Gastrophryne olivacea
Eastern: G. carolinensis
Sheep: Hypopachus variolosus
137 Giant: Dicamptodon ensatus
Chuckwalla: Sauromalus obesus
138 Necturus maculosus
139 Two-toed Amphiuma: Amphiuma means
Hellbender: Cryptobranchus alleganiensis
140 Greater: Siren lacertina
Dwarf: Pseudobranchus striatus
141 Pacific Giant: Dicamptodon ensatus
Olympic: Rhyacotriton olympicus
142 Notophthalmus viridescens viridescens
143 Eastern: Notophthalmus viridescens viridescens
Pacific: Taricha torosa
145 Spotted: Ambystoma maculatum
Tiger: A. tigrinum
Marbled: A. opacum
Jefferson: A. jeffersonianum
Smallmouth: A. texanum
146 Dusky: Desmognathus fuscus
Mountain Dusky: D. ochrophaeus
147 Redback: Plethodon cinereus
Slimy: P. glutinosus
148 Monterey Ensatina: Ensatina eschscholtzii eschscholtzii
Large-blotched Ensatina: E. eschscholtzii klauberi
California Slender: Batrachoseps attenuatus
149 Arboreal: Aneides lugubris
Green: A. aeneus
150 Grotto: Typhlotriton spelaeus
Texas Blind: Typhlomolge rathbuni
151 Red: Pseudotriton ruber
Spring: Gyrinophilus porphyriticus
152 Two-lined: Eurycea bislineata
Longtail: E. longicauda
Cave: E. lucifuga
153 Hemidactylium scutatum

INDEX

An asterisk (*) designates pages that are illustrated; **bold type** denotes pages containing more extensive information.

MEASURING SCALE (IN MILLIMETERS AND CENTIMETERS)

K L